Ben Stacy Jerrik (Ed.)

Coverage (Telecommunication)

Ben Stacy Jerrik (Ed.)

Coverage (Telecommunication)

Broadcasting, Radio broadcasting, Telecommunication

Part Press

Contents

Coverage_(telecommunication)

In telecommunications, the **coverage** of a radio station is the geographic area where the station can communicate. Broadcasters and telecommunications companies frequently produce coverage maps to indicate to users the station's intended service area. Coverage depends on several factors, such as orography (i.e. mountains) and buildings, technology and radio frequency. Some frequencies provide better regional coverage, while other frequencies penetrate better through obstacles, such as buildings in cities.

The ability of a mobile phone to connect to a base station depends on the strength of the signal. That may be boosted by higher power transmissions, better antennae and taller antenna masts. Signals will also need to be boosted to pass through buildings, which is a particular problem designing networks for large metropolitan areas with modern skyscrapers. Signals also do not travel deep underground, so specialized transmission solutions are used to deliver mobile phone coverage into areas such as underground parking garages and subway trains.

Coverage noticer

A coverage noticer is a device that beeps (or vibrates) when in a zone that lacks coverage (white spot). This is fundamental for critical services (security, emergency and so on). When the user goes to a covered area, the noticer ceases beeping.

It can be integrated in a mobile phone also.

See also

- Femtocell
- Footprint (satellite)
- Indoor and outdoor
- Mobile phone
- Mobile computing
- Roaming
- Telecommunications network
- White spot
- Wireless

External links

- Understanding Cell Phone Coverage Areas. FCC Consumer Facts [1].
- CRC-COVWEB - A free online program that calculates radio wave coverage [2].
- Wi-Fi coverage [3]PDF (32.9 KiB)

References

[1] http://www.state.sd.us/puc/telecom/coveragearea.htm
[2] http://lrcov.crc.ca
[3] http://www.itaa.org/isec/pubs/e20063-05.pdf

Broadcasting

Broadcasting is the distribution of audio and video content to a dispersed audience via any audio visual medium. Receiving parties may include the general public or a relatively large subset thereof. It has been used for purposes of private recreation, non-commercial exchange of messages, experimentation, self-training, and emergency communication such as amateur (ham) radio and amateur television (ATV) in addition to commercial purposes like popular radio or TV stations with advertisements.

Broadcasting antenna in Stuttgart

History

The term *broadcast* was first adopted by early radio engineers from the Midwestern United States,[1] treating broadcast sowing as a metaphor for the dispersal inherent in omnidirectional radio signals.Broadcasting is a very large and significant segment of the mass media.

Originally all broadcasting was composed of analog signals using analog transmission techniques and more recently broadcasters have switched to digital signals using digital transmission.

- Analog audio vs. HD Radio
- Analog television vs. Digital television
- Wireless

The world's technological capacity to receive information through one-way broadcast networks more than quadrupled during the two decades from 1986 to 2007, from 432 exabytes of (optimally compressed) information, to 1.9 zettabytes.[2] This is the information equivalent of 55 newspapers per person per day in 1986, and 175 newspapers per person per day by 2007.[3]

Types of electronic broadcasting

Historically, there have been several different types of electronic media broadcasting:

- Telephone broadcasting (1881–1932): the earliest form of electronic broadcasting (not counting data services offered by stock telegraph companies from 1867, if ticker-tapes are excluded from the definition). Telephone broadcasting began with the advent of Théâtrophone ("Theatre Phone") systems, which were telephone-based distribution systems allowing subscribers to listen to live opera and theatre performances over telephone lines, created by French inventor Clément Ader in 1881. Telephone broadcasting also grew to include telephone newspaper services for news and entertainment programming which were introduced in the 1890s, primarily located in large European cities. These telephone-based subscription services were the first examples of electrical/electronic broadcasting and offered a wide variety of programming.
- Radio broadcasting (experimentally from 1906, commercially from 1920): radio broadcasting is an audio (sound) broadcasting service, broadcast through the air as radio waves from a transmitter to an radio antenna and, thus, to a receiver. Stations can be linked in radio networks to broadcast common radio programs, either in broadcast syndication, simulcast or subchannels.
- History of television broadcasting (telecast), experimentally from 1925, commercial television from the 1930s: this television programming medium was long-awaited by the general public and rapidly rose to compete with its older radio-broadcasting sibling.
- Cable radio (also called "cable FM", from 1928) and cable television (from 1932): both via coaxial cable, serving principally as transmission mediums for programming produced at either radio or television stations, with limited

production of cable-dedicated programming.

- Direct-broadcast satellite (DBS) (from circa 1974) and satellite radio (from circa 1990): meant for direct-to-home broadcast programming (as opposed to studio network uplinks and downlinks), provides a mix of traditional radio or television broadcast programming, or both, with dedicated satellite radio programming. (See also: Satellite television)
- Webcasting of video/television (from circa 1993) and audio/radio (from circa 1994) streams: offers a mix of traditional radio and television station broadcast programming with dedicated internet radio-webcast programming.

Economic models

Economically there are a few ways in which stations are able to broadcast continually. Each differs in the method by which stations are funded:

- in-kind donations of time and skills by volunteers (common with community radio broadcasters)
- direct government payments or operation of public broadcasters
- indirect government payments, such as radio and television licenses
- grants from foundations or business entities
- selling advertising or sponsorships
- public subscription or membership

Broadcasters may rely on a combination of these business models. For example, National Public Radio (NPR), a non-commercial educational (NCE) public radio media organization within the U.S., receives grants from the Corporation for Public Broadcasting (CPB) (which, in turn, receives funding from the U.S. government), by public membership and by selling "extended credits" to corporations.

Underwriting spots vs. commercials

In contrast with commercial broadcasting, NPR does not carry traditional radio commercials or television commercial. It offers major donors brief statements that are called underwriting spots and unlike commercials, are governed by specific FCC restrictions in addition to the truth-in-advertising laws; they cannot advocate a product or contain any "call to action"

Recorded broadcasts and live broadcasts

The first regular television broadcasts began in 1937. Broadcasts can be classified as "recorded" or "live". The former allows correcting errors, and removing superfluous or undesired material, rearranging it, applying slow-motion and repetitions, and other techniques to enhance the program. However, some live events like sports television can include some of the aspects including slow-motion clips of important goals/hits, etc., in between the live television telecast.

A television studio production control room in Olympia, Washington, August 2008.

American radio-network broadcasters habitually forbade prerecorded broadcasts in the 1930s and 1940s requiring radio programs played for the Eastern and Central time zones to be repeated three hours later for the Pacific time zone (See: Effects of time on North American broadcasting). This restriction was dropped for special occasions, as in the case of the German dirigible airship *Hindenburg* disaster at Lakehurst, New Jersey, in 1937. During World War II, prerecorded broadcasts from war correspondents were allowed on U.S. radio. In addition, American radio programs were recorded for playback by Armed Forces Radio radio stations around the world.

A disadvantage of recording first is that the public may know the outcome of an event from another source, which may be a "spoiler". In addition, prerecording prevents live radio announcers from deviating from an officially approved script, as occurred with propaganda broadcasts from Germany in the 1940s and with Radio Moscow in the 1980s.

Many events are advertised as being live, although they are often "recorded live" (sometimes called "live-to-tape"). This is particularly true of performances of musical artists on radio when they visit for an in-studio concert performance. Similar situations have occurred in television production (*The Cosby Show* is recorded in front of a live television studio audience") and news broadcasting.

A broadcast may be distributed through several physical means. If coming directly from the radio studio at a single station or television station, it is simply sent through the studio/transmitter link to the transmitter and thence from the television antenna located on the radio masts and towers out to the world. Programming may also come through a communications satellite, played either live or recorded for later transmission. Networks of stations may simulcast the same programming at the same time, originally *via* microwave link, now usually by satellite.

Distribution to stations or networks may also be through physical media, such as magnetic tape, compact disc (CD), DVD, and sometimes other formats. Usually these are included in another broadcast, such as when electronic news gathering (ENG) returns a story to the station for inclusion on a news programme.

The final leg of broadcast distribution is how the signal gets to the listener or viewer. It may come over the air as with a radio station or television station to an antenna and radio receiver, or may come through cable television [4] or cable radio (or "wireless cable") *via* the station or directly from a network. The Internet may also bring either internet radio or streaming media television to the recipient, especially with multicasting allowing the signal and bandwidth to be shared.

The term "broadcast network" is often used to distinguish networks that broadcast an over-the-air television signals that can be received using a tuner (television) inside a television set with a television antenna from so-called networks that are broadcast only *via* cable television (cablecast) or satellite television that uses a dish antenna. The term "broadcast television" can refer to the television programs of such networks.

Social impact

Radio station WTUL studio, Tulane University, New Orleans

The sequencing of content in a broadcast is called a schedule. As with all technological endeavours, a number of technical terms and slang have developed. A list of these terms can be found at List of broadcasting terms. Television and radio programs are distributed through radio broadcasting or cable, often both simultaneously. By coding signals and having a cable converter box with decoding equipment in homes, the latter also enables subscription-based channels, pay-tv and pay-per-view services.

In his essay, John Durham Peters wrote that communication is a tool used for dissemination. Durham stated, "Dissemination is a lens- sometimes a usefully distorting one- that helps us tackle basic issues such as interaction, presence, and space and time...on the agenda of any future communication theory in general" (Durham, 211). Dissemination focuses on the message being relayed from one main source to one large audience without the exchange of dialogue in between. There's chance for the message to be tweaked or corrupted once the main source releases it. There is really no way to predetermine how the larger population or audience will absorb the message. They can choose to listen, analyze, or simply ignore it. Dissemination in communication is widely used in the world of broadcasting.

Broadcasting focuses on getting one message out and it is up to the general public to do what they wish with it. Durham also states that broadcasting is used to address an open ended destination (Durham, 212). There are many forms of broadcast, but they all aim to distribute a signal that will reach the target audience. Broadcasting can arrange audiences into entire assemblies (Durham, 213).

In terms of media broadcasting, a radio show can gather a large number of followers who tune in every day to specifically listen to that specific disc jockey. The disc jockey follows the script for his or her radio show and just talks into the microphone. He or she does not expect immediate feedback from any listeners. The message is broadcast across airwaves throughout the community, but there the listeners cannot always respond immediately, especially since many radio shows are recorded prior to the actual air time.

See also

- 1worldspace – world's first commercial satellite radio direct-to-home broadcaster
- Analog television
- Bandplan
- Broadcast quality
- Broadcast television systems – contains the standards of the topic
- Broadcasting in the United States
- Cablecast
- Dead air
- Digital television
- Electronic media
- European Broadcasting Union (EBU)
- List of broadcasting terms
- List of broadcast satellites
- NaSTA
- Nonbroadcast Multiple Access Network (NBMA)
- North American broadcast television frequencies
- Outside broadcast
- Radio Act of 1927
- Reality television
- Society of Broadcast Engineers (SBE)
- Television broadcasting in Australia
- Television transmitter
- Transposer

Notes

[1] http://www.ovguide.com/broadcasting-9202a8c04000641f80000000000c758b#

[2] "The World's Technological Capacity to Store, Communicate, and Compute Information" (http://www.sciencemag.org/content/332/6025/ 60), Martin Hilbert and Priscila López (2011), Science (journal), 332(6025), 60-65; free access to the article through here: martinhilbert.net/WorldInfoCapacity.html

[3] "video animation on The World's Technological Capacity to Store, Communicate, and Compute Information from 1986 to 2010" (http:// ideas.economist.com/video/giant-sifting-sound-0). Ideas.economist.com. . Retrieved 2011-12-26.

[4] http://www.diwaxx.ru/

Bibliography

- Carey, James (1989) *Communication as Culture*, Routledge, New York and London, pp. 201–30
- Kahn, Frank J., ed. *Documents of American Broadcasting,* fourth edition (Prentice-Hall, Inc., 1984).
- Lichty Lawrence W., and Topping Malachi C., eds. *American Broadcasting: A Source Book on the History of Radio and Television* (Hastings House, 1975).
- Meyrowitz, Joshua., *Mediating Communication: What Happens?* in Downing, J., Mohammadi, A., and Sreberny-Mohammadi, A., (eds) *Questioning The Media* (Sage, Thousand Oaks, 1995) pp. 39–53
- Peters, John Durham. "Communication as Dissemination." Communication as…Perspectives on Theory. Thousand Oakes, CA: Sage, 2006. 211-22.
- Thompson, J., *The Media and Modernity,* in Mackay, H and O'Sullivan, T (eds) *The Media Reader: Continuity and Transformation.*, (Sage, London, 1999) pp. 12–27

Further reading

- Gilbert, Sean; Nelson, John; Jacobs, George, *World Radio TV Handbook 2007* (http://books.google.com/books?id=IBu8NHvC4fMC&printsec=frontcover), Watson-Guptill, 2006. ISBN 0-9535864-9-9. The 2007 edition of the *World Radio TV Handbook.*
- Wells, Alan, *World Broadcasting: A Comparative View* (http://books.google.com/books?id=3zpeKLHPVBQC&printsec=frontcover), Greenwood Publishing Group, 1996. ISBN 1-56750-245-8

External links

- Radio Locator (http://www.radio-locator.com), for American radio station with format, power, and coverage information.
- Jim Hawkins' Radio and Broadcast Technology Page (http://www.hawkins.pair.com/radio.html) – History of broadcast transmitter technology

Radio_broadcasting

Radio broadcasting is a one-way wireless transmission over radio waves intended to reach a wide audience. Stations can be linked in radio networks to broadcast a common radio format, either in broadcast syndication or simulcast or both. Audio broadcasting also can be done via cable radio, local wire television networks, satellite radio, and internet radio via streaming media on the Internet.

The signal types can be either analog audio or digital audio.

Amateur radio (also, **ham radio**) is a form of radio communication that is the private use of designated radio bands, for purposes of private recreation, non-commercial exchange of messages, experimentation, self-training, and emergency communication.[1]

Long wave radio broadcasting station, Motala, Sweden

History

The earliest radio stations were simply radiotelegraphy systems and did not carry audio. The first claimed audio transmission that could be termed a *broadcast* occurred on Christmas Eve in 1906, and was made by Reginald Fessenden. Whether this broadcast actually took place is disputed.[2] While many early experimenters attempted to create systems similar to radiotelephone devices by which only two parties were meant to communicate, there were others who intended to transmit to larger audiences. Charles Herrold started broadcasting in California in 1909 and was carrying audio by the next year. (Herrold's station eventually became KCBS).

Broadcasting tower in Trondheim, Norway

For the next decade, radio tinkerers had to build their own radio receivers. In The Hague, the Netherlands, PCGG started broadcasting on November 6, 1919. In 1916, Frank Conrad, an employee for the Westinghouse Electric Corporation, began broadcasting from his Wilkinsburg, Pennsylvania garage with the call letters 8XK. Later, the station was moved to the top of the Westinghouse factory building in East Pittsburgh, Pennsylvania. Westinghouse relaunched the station as KDKA on November 2, 1920, claiming to be "the world's first commercially licensed radio station".[3] The commercial broadcasting designation came from the type of broadcast license; advertisements did not air until years later. The first licensed broadcast in the United States came from KDKA itself: the results of the Harding/Cox Presidential Election. The Montreal station that became CFCF began broadcast programming on May 20, 1920, and the Detroit station that became WWJ began program broadcasts beginning on August 20, 1920, although neither held a license at the time.

Radio Argentina began regularly scheduled transmissions from the Teatro Coliseo in Buenos Aires on August 27, 1920, making its own priority claim. The station got its license on November 19, 1923. The delay was due to the lack of official Argentine licensing procedures before that date. This station continued regular broadcasting of entertainment and cultural fare for several decades.[4]

Radio in education soon followed and colleges across the U.S. began adding radio broadcasting courses to their curricula. Curry College in Milton, Massachusetts introduced one of the first broadcasting majors in 1932 when the college teamed up with WLOE in Boston to have students broadcast programs.[5]

Types

Broadcasting by radio takes several forms. These include AM and FM stations. There are several subtypes, namely commercial broadcasting, non-commercial educational (NCE) public broadcasting and non-profit varieties as well as community radio, student-run campus radio stations and hospital radio stations can be found throughout the world.

Many stations broadcast on shortwave bands using AM technology that can be received over thousands of miles (especially at night). For example, the BBC, VOA, VOR, and Deutsche Welle have transmitted

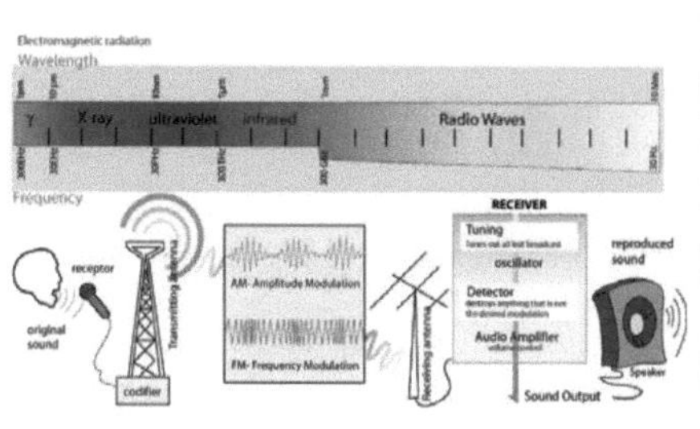

Transmission and reception schematic

via shortwave to Africa and Asia. These broadcasts are very sensitive to atmospheric conditions and solar activity.

Arbitron, the United States-based company that reports on radio audiences, defines a "radio station" as a government-licensed AM or FM station; an HD Radio (primary or multicast) station; an internet stream of an existing government-licensed station; one of the satellite radio channels from XM Satellite Radio or Sirius Satellite Radio; or, potentially, a station that is not government licensed.[6]

Shortwave

See Shortwave for the differences between shortwave, medium wave and long wave spectra. Used largely for national broadcasters, international propaganda, or religious broadcasting organizations.

AM

AM stations were the earliest broadcasting stations to be developed. AM refers to amplitude modulation, a mode of broadcasting radio waves by varying the amplitude of the carrier signal in response to the amplitude of the signal to be transmitted.

The medium-wave band is used worldwide for AM broadcasting. Europe also uses the long wave band. In response to the growing popularity of FM radio stereo radio stations in the late 1980s and early

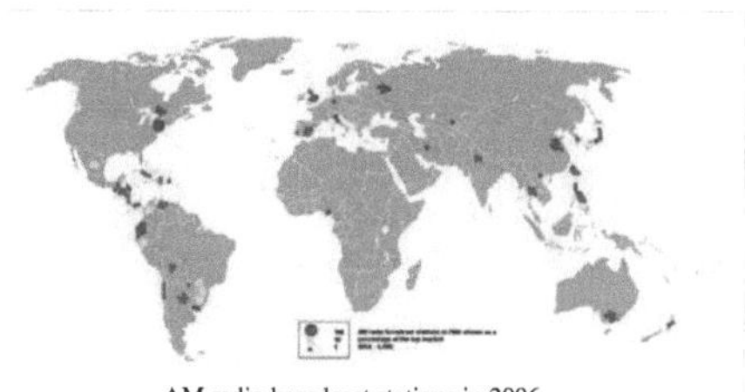
AM radio broadcast stations in 2006

1990s, some North American stations began broadcasting in AM stereo, though this never gained popularity, and very few receivers were ever sold.

One of the advantages of AM is that its signal can be detected (turned into sound) with simple equipment. If a signal is strong enough, not even a power source is needed; building an unpowered crystal radio receiver was a common childhood project in the early decades of AM broadcasting.

AM broadcasts occur on North American airwaves in the medium wave frequency range of 530 to 1700 kHz (known as the "standard broadcast band"). The band was expanded in the 1990s by adding nine channels from 1620 to 1700 kHz. Channels are spaced every 10 kHz in the Americas, and generally every 9 kHz everywhere else.

The signal is subject to interference from electrical storms (lightning) and other electromagnetic interference (EMI).

AM transmissions cannot be ionospherically propagated during the day due to strong absorption in the D-layer of the ionosphere. In a crowded channel environment this means that the power of regional channels which share a frequency must be reduced at night or directionally beamed in order to avoid interference, which reduces the potential nighttime audience. Some stations have frequencies unshared with other stations in North America; these are called clear-channel stations. Many of them can be heard across much of the country at night. This is not to be confused with Clear Channel Communications, merely a brand name, which currently owns many U.S. radio stations on both the AM and FM bands. During the night, this absorption largely disappears and permits signals to travel to much more distant locations via ionospheric reflections. However, fading of the signal can be severe at night.

AM radio transmitters can transmit audio frequencies up to 15 kHz (now limited to 10 kHz in the US due to FCC rules designed to reduce interference), but most receivers are only capable of reproducing frequencies up to 5 kHz or less. At the time that AM broadcasting began in the 1920s, this provided adequate fidelity for existing microphones, 78 rpm recordings, and loudspeakers. The fidelity of sound equipment subsequently improved considerably, but the receivers did not. Reducing the bandwidth of the receivers reduces the cost of manufacturing and makes them less prone to interference. AM stations are never assigned adjacent channels in the same service area. This prevents the sideband power generated by two stations from interfering with each other.[7] Bob Carver created an AM stereo tuner employing notch filtering that demonstrated that an AM broadcast can meet or exceed the 15 kHz baseband bandwidth alloted to FM stations without objectionable interference. After several years, the tuner was discontinued. Bob Carver had left the company and the Carver Corporation later cut the number of models produced before discontinuing production completely.

FM

FM refers to frequency modulation, and occurs on VHF airwaves in the frequency range of 88 to 108 MHz everywhere (except Japan and Russia). Japan uses the 76 to 90 MHz band. Russia has two bands widely used by the Soviet Union, 65.9 to 74 MHz and 87.5 to 108 MHz worldwide standard. FM stations are much more popular since higher sound fidelity and stereo broadcasting became common in this format.

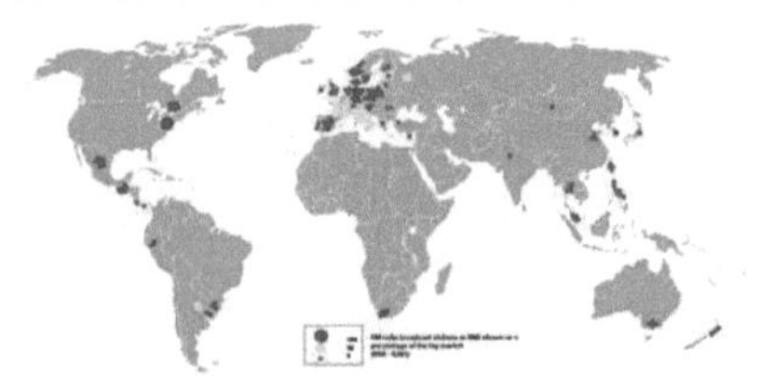
FM radio broadcast stations in 2006

FM radio was invented by Edwin H. Armstrong in the 1930s for the specific purpose of overcoming the interference problem of AM radio, to which it is relatively immune. At the same time, greater fidelity was made possible by spacing stations further apart. Instead of 10 kHz apart, as on the AM band in the US, FM channels are 200 kHz (0.2 MHz) apart. In other countries greater spacing is sometimes mandatory, such as in New Zealand, which uses 700 kHz spacing (previously 800 kHz). The improved fidelity made available was far in advance of the audio equipment of the 1940s, but wide interchannel spacing was chosen to take advantage of the noise-suppressing feature of wideband FM.

Bandwidth of 200 kHz is not needed to accommodate an audio signal — 20 kHz to 30 kHz is all that is necessary for a narrowband FM signal. The 200 kHz bandwidth allowed room for ±75 kHz signal deviation from the assigned frequency, plus guard bands to reduce or eliminate adjacent channel interference. The larger bandwidth allows for broadcasting a 15 kHz bandwidth audio signal plus a 38 kHz stereo "subcarrier"—a piggyback signal that rides on the main signal. Additional unused capacity is used by some broadcasters to transmit utility functions such as background music for public areas, GPS auxiliary signals, or financial market data.

The AM radio problem of interference at night was addressed in a different way. At the time FM was set up, the available frequencies were far higher in the spectrum than those used for AM radio - by a factor of approximately 100. Using these frequencies meant that even at far higher power, the range of a given FM signal was much shorter; thus its market was more local than for AM radio. The reception range at night is the same as in the daytime.

The original FM radio service in the U.S. was the Yankee Network, located in New England.[8] [9] [10] Regular FM broadcasting began in 1939, but did not pose a significant threat to the AM broadcasting industry. It required purchase of a special receiver. The frequencies used, 42 to 50 MHz, were not those used today. The change to the current frequencies, 88 to 108 MHz, began after the end of World War II, and was to some extent imposed by AM broadcasters as an attempt to cripple what was by now realized to be a potentially serious threat.

FM radio on the new band had to begin from the ground floor. As a commercial venture it remained a little-used audio enthusiasts' medium until the 1960s. The more prosperous AM stations, or their owners, acquired FM licenses and often broadcast the same programming on the FM station as on the AM station ("simulcasting"). The FCC limited this practice in the 1970s. By the 1980s, since almost all new radios included both AM and FM tuners, FM became the dominant medium, especially in cities. Because of its greater range, AM remained more common in rural environments.

Pirate radio

Pirate radio is radio broadcasting not sanctioned by the regulations of the originating country. Pirate radio may be a commercial enterprise supported by advertising targeted to listeners in the reception area, or may be privately run for entertainment, or political reasons, sometimes on a very small scale covering only a few city blocks.

Terrestrial digital radio

Digital radio broadcasting has emerged, first in Europe (the UK in 1995 and Germany in 1999), and later in the United States, France, the Netherlands, South Africa and many other countries worldwide. The most simple system is named DAB Digital Radio, for Digital Audio Broadcasting, and uses the public domain EUREKA 147 (Band III) system. DAB is used mainly in the UK and South Africa. Germany and Holland use the DAB and DAB+ systems, and France uses the L-Band system of DAB Digital Radio.

In the United States, digital radio isn't used in the same way as Europe and South Africa. Instead, the IBOC system is named HD Radio and owned by a consortium of private companies that is called iBiquity. An international non-profit consortium Digital Radio Mondiale (DRM), has introduced the public domain DRM system.

Satellite

Satellite radio broadcasters are slowly emerging, but the enormous entry costs of space-based satellite transmitters, and restrictions on available radio spectrum licenses has restricted growth of this market. In the USA and Canada, just two services, XM Satellite Radio and Sirius Satellite Radio exist. Both XM and Sirius are owned by Sirius XM Radio, which was formed by the merger of XM and Sirius on July 29, 2008, whereas in Canada, XM Radio Canada and Sirius Canada remain separate companies.

Program formats

Radio program formats differ by country, regulation and markets. For instance, the U.S. Federal Communications Commission designates the 88–92 megahertz band in the U.S. for non-profit or educational programming, with advertising prohibited.

In addition, formats change in popularity as time passes and technology improves. Early radio equipment only allowed program material to be broadcast in real time, known as *live* broadcasting. As technology for sound recording improved, an increasing proportion of broadcast programming used pre-recorded material. A current trend is the automation of radio stations. Some stations now operate without direct human intervention by using entirely pre-recorded material sequenced by computer control.

See also

- Call sign
- Construction permit
- Disc jockey (DJ)
- History of broadcasting
- International broadcasting
- List of radio topics
- Low power radio station
- Radio
- Radio antenna
- Radio network
- Radio personality
- RF modulation
- Sports commentator
- Television station

References

[1] "Ham Radio Frequently Asked Questions" (http://www.arrl.org/ham-radio-faq-s). *ARRL.org.* . Retrieved 2010-05-23.

[2] Fessenden — The Next Chapter RWonline.com (http://www.rwonline.com/article/72046)

[3] Baudino, Joseph E; John M. Kittross (Winter, 1977). "Broadcasting's Oldest Stations: An Examination of Four Claimants" (http://www. ieee.org/web/aboutus/history_center/kdka.html). *Journal of Broadcasting*: pp. 61–82. . Retrieved 2008-08-08.

[4] Atgelt, Carlos A. "Early History of Radio Broadcasting in Argentina." (http://www.oldradio.com/archives/international/argentin.html) The Broadcast Archive (Oldradio.com).

[5] www.curry.edu

[6] "What is a Radio Station?" (http://www.nxtbook.com/nxtbooks/newbay/rw_20081008/index.php). *Radio World*: p. 6. .

[7] http://kwarner.bravehost.com/tech.htm

[8] Halper, Donna L. "John Shepard's FM Stations—America's first FM network." (http://www.bostonradio.org/essays/shepard-fm.html) Boston Radio Archives (BostonRadio.org).

[9] "The Yankee Network in 1936." (http://www.bostonradio.org/yankee-36.html) Boston Radio Archives (BostonRadio.org)

[10] Miller, Jeff. "FM Broadcasting Chronology." (http://jeff560.tripod.com/chronofm.html) Rev. 2005-12-27.

Further reading

- Briggs Asa. *The History of Broadcasting in the United Kingdom* (Oxford University Press, 1961).
- Ewbank Henry and Lawton Sherman P. *Broadcasting: Radio and Television* (Harper & Brothers, 1952).
- Fisher, Marc *Something In The Air: Radio, Rock, and the Revolution That Shaped A Generation* (Random House, 2007).
- Lewis, Tom, *Empire of the Air: The Men Who Made Radio*, 1st ed., New York : E. Burlingame Books, 1991. ISBN 0-06-018215-6. "Empire of the Air: The Men Who Made Radio" (1992) by Ken Burns was a PBS documentary based on the book.
- Ray, William B. *FCC: The Ups and Downs of Radio-TV Regulation* (Iowa State University Press, 1990).
- Russo, Alexan der. *Points on the Dial: Golden Age Radio Beyond the Networks* (Duke University Press; 2010) 278 pages; discusses regional and local radio as forms that "complicate" the image of the medium as a national unifier from the 1920s to the 1950s.
- Scannell, Paddy, and Cardiff, David. *A Social History of British Broadcasting, Volume One, 1922-1939* (Basil Blackwell, 1991).
- Schwoch James. *The American Radio Industry and Its Latin American Activities, 1900-1939* (University of Illinois Press, 1990).
- White Llewellyn. *The American Radio* (University of Chicago Press, 1947).

External links

Patents

- **U.S. Patent 1082221** (http://www.google.com/patents?vid=1082221), Georg Graf von Arco, "Radiotelegraphic station" (December 1913)
- **U.S. Patent 1116111** (http://www.google.com/patents?vid=1116111), Richard Pfund, "Station for the transmission and reception of electromagnetic wave energy". (November 1914)
- **U.S. Patent 1214591** (http://www.google.com/patents?vid=1214591), Gustav Reuthe, "Antenna for radiotelegraph station" (February 1917)

General

- Federal Communications Commission website (http://www.fcc.gov/), fcc.gov
- DXing.info (http://www.dxing.info) - Information about radio stations worldwide
- Radio-Locator.com (http://www.radio-locator.com/)- Links to 10,000 radio stations worldwide.
- BBC reception advice (http://www.bbc.co.uk/reception/)
- DXradio.50webs.com (http://dxradio.50webs.com) "The SWDXER" - with general SWL information and radio antenna tips.
- RadioStationZone.com (http://www.radiostationzone.com) - 10.000+ radio stations worldwide with ratings, comments and listen live links
- RadioBeta.com (http://www.radiobeta.com), search for stations around the globe
- Online-Radio-Stations.org (http://www.online-radio-stations.org) - The Web Radio Tuner has a comprehensive list of over 50.000 radio stations
- RadioStations.com (http://www.radiostations.com/) has a directory of radio stations and real-time music listings
- ZoZanga.com (http://www.zozanga.com/internetstuff/englishradiostations.htm) List of radio stations and real-time music listings
- MyTravelTunes.com (http://www.mytraveltunes.com/) Search for radio stations throughout the U.S.
- UnwantedEmissions.com (http://www.unwantedemissions.com/) - A general reference to radio spectrum allocations.
- Radio stanice (http://www.navidiku.rs/radio-stanice/) - Search for radio stations throughout the Europe
- NEC Lab (http://www.ingenierias.ugto.mx/profesores/sledesma/documentos/index.htm) - A tool to design and test antennas for Radio broadcasting.

Telecommunication

Telecommunication is the transmission of information over significant distances to communicate. In earlier times, telecommunications involved the use of visual signals, such as beacons, smoke signals, semaphore telegraphs, signal flags, and optical heliographs, or audio messages via coded drumbeats, lung-blown horns, or sent by loud whistles, for example. In the modern age of electricity and electronics, telecommunications now also includes the use of electrical devices such as the telegraph, telephone, and teleprinter, as well as the use of radio and microwave communications, as well as fiber optics and their associated electronics, plus the use of the orbiting satellites and the Internet.

A parabolic satellite communication antenna at the biggest facility for satellite communication in Raisting, Bavaria, Germany

A revolution in wireless telecommunications began in the first decade of the 20th century with pioneering developments in wireless radio communications by Nikola Tesla and Guglielmo Marconi. Marconi won the Nobel Prize in Physics in 1909 for his efforts. Other highly notable pioneering inventors and developers in the field of electrical and electronic telecommunications include Charles Wheatstone and Samuel Morse (telegraph), Alexander Graham Bell (telephone), Edwin Armstrong, and Lee de Forest (radio), as well as John Logie Baird and Philo Farnsworth (television).

The world's effective capacity to exchange information through two-way telecommunication networks grew from 281 petabytes of (optimally compressed) information in 1986, to 471 petabytes in 1993, to 2.2 (optimally compressed) exabytes in 2000, and to 65 (optimally compressed) exabytes in 2007.[1] This is the informational equivalent of 2 newspaper pages per person per day in 1986, and 6 entire newspapers per person per day by 2007.[2] Given this growth, telecommunications play an increasingly important role in the world

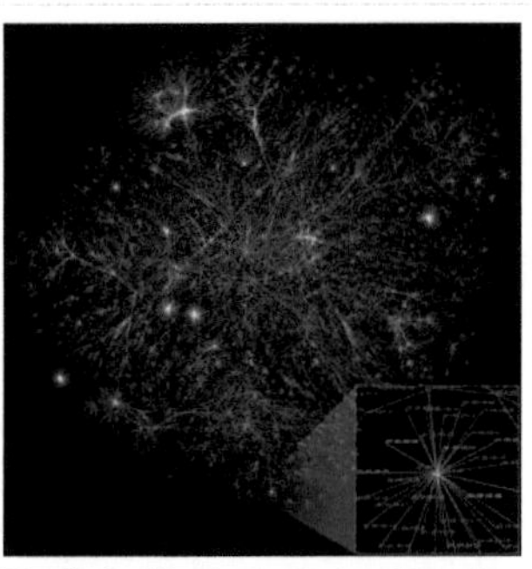

Visualization from the Opte Project of the various routes through a portion of the Internet

economy and the worldwide telecommunication industry's revenue was estimated to be $3.85 trillion in 2008.[3] The service revenue of the global telecommunications industry was estimated to be $1.7 trillion in 2008, and is expected to touch $2.7 trillion by 2013.[3]

Etymology

The word *telecommunication* was adapted from the French word *télécommunication*. It is a compound of the Greek prefix *tele-* (τηλε-), meaning "far off", and the Latin *communicare*, meaning "to share".[4] The French word *télécommunication* was first invented in the French Grande Ecole "Telecom ParisTech" formerly known as "Ecole nationale superieure des telecommunications" in 1904 by the French engineer and novelist Édouard Estaunié.[5]

History

Ancient systems

Greek hydraulic semaphore systems were used as early as the 4th century BC. The hydraulic semaphores, which worked with water filled vessels and visual signals, functioned as optical telegraphs. However, they could only utilize a very limited range of pre-determined messages, and as with all such optical telegraphs could only be deployed during good visibility conditions.[6]

During the Middle Ages, chains of beacons were commonly used on hilltops as a means of relaying a signal. Beacon chains suffered the drawback that they could only pass a single bit of information, so the meaning of the message such as "the enemy has been sighted" had to be agreed upon in advance. One notable instance of their use was during the Spanish Armada, when a beacon chain relayed a signal from Plymouth to London that signaled the arrival of the Spanish warships.[7]

Systems since the Middle Ages

In 1792, Claude Chappe, a French engineer, built the first fixed visual telegraphy system (or semaphore line) between Lille and Paris.[8] However semaphore systems suffered from the need for skilled operators and the expensive towers at intervals of 10–30 kilometers (6–20 mi). As a result of competition from the electrical telegraph, Europe's last commercial semaphore line in Sweden was abandoned in 1880.[9]

Telegraph and telephone

The first commercial electrical telegraph was constructed by Sir Charles Wheatstone and Sir William Fothergill Cooke, and its use began on April 9, 1839. Both Wheatstone and Cooke viewed their device as "an improvement to the [already-existing, so-called] electromagnetic telegraph" not as a new device.[10]

The businessman Samuel F.B. Morse and the physicist Joseph Henry of the United States developed their own, simpler version of the electrical telegraph, independently. Morse successfully demonstrated

A replica of one of Chappe's semaphore towers in Nalbach

this system on September 2, 1837. Morse's most important technical contribution to this telegraph was the rather simple and highly efficient Morse Code, which was an important advance over Wheatstone's complicated and significantly more expensive telegraph system. The communications efficiency of the Morse Code anticipated that of the Huffman code in digital communications by over 100 years, but Morse and his associate Alfred Vail developed the code purely empirically, unlike Huffman, who gave a detailed theoretical explanation of how his method worked.

The first permanent transatlantic telegraph cable was successfully completed on 27 July 1866, allowing transatlantic electrical communication for the first time.[11] An earlier transatlantic cable had operated for a few months in 1859, and among other things, it carried messages of greeting back and forth between President James Buchanan of the United States and Queen Victoria of the United Kingdom.

However, that transatlantic cable failed soon, and the project to lay a replacement line was delayed for five years by the American Civil War. Also, these transatlantic cables would have been completely incapable of carrying telephone calls even had the telephone already been invented. The first transatlantic telephone cable (which incorporated hundreds of electronic amplifiers) was not operational until 1956.[12]

The conventional telephone now in use worldwide was first patented by Alexander Graham Bell in March 1876.[13] That first patent by Bell was the *master patent* of the telephone, from which all other patents for electric telephone devices and features flowed. Credit for the invention of the electric telephone has been frequently disputed, and new controversies over the issue have arisen from time-to-time. As with other great inventions such as radio, television, the light bulb, and the digital computer, there were several inventors who did pioneering experimental work on *voice transmission over a wire*, and then they improved on each other's ideas. However, the key innovators were Alexander Graham Bell and Gardiner Greene Hubbard, who created the first telephone company, the Bell Telephone Company in the United States, which later evolved into American Telephone & Telegraph (AT&T).

The first commercial telephone services were set up in 1878 and 1879 on both sides of the Atlantic in the cities of New Haven, Connecticut, and London, England.[14] [15]

Radio and television

In 1832, James Lindsay gave a classroom demonstration of wireless telegraphy via conductive water to his students. By 1854, he was able to demonstrate a transmission across the Firth of Tay from Dundee, Scotland, to Woodhaven, a distance of about two miles (3 km), again using water as the transmission medium.[16] In December 1901, Guglielmo Marconi established wireless communication between St. John's, Newfoundland and Poldhu, Cornwall (England), earning him the Nobel Prize in Physics for 1909, one which he shared with Karl Braun.[17] However *small-scale* radio communication had already been demonstrated in 1893 by Nikola Tesla in a presentation before the National Electric Light Association.[18]

On March 25, 1925, John Logie Baird of Scotland was able to demonstrate the transmission of moving pictures at the Selfridge's department store in London, England. Baird's system relied upon the fast-rotating Nipkow disk, and thus it became known as the mechanical television. It formed the basis of experimental broadcasts done by the British Broadcasting Corporation beginning September 30, 1929.[19] However, for most of the 20th century, television systems were designed around the cathode ray tube, invented by Karl Braun. The first version of such an electronic television to show promise was produced by Philo Farnsworth of the United States, and it was demonstrated to his family in Idaho on September 7, 1927.[20]

Television, however, is not solely a technology, limited to its basic and practical application. It functions both as an appliance, and also as a means for social story telling and message dissemination. It is a cultural tool that provides a communal experience of receiving information and experiencing fantasy. It acts as a "window to the world" by bridging audiences from all over through programming of stories, triumphs, and tragedies that are outside of personal experiences. [21]

Computer networks and the Internet

On 11 September 1940, George Stibitz was able to transmit problems using teleprinter to his Complex Number Calculator in New York and receive the computed results back at Dartmouth College in New Hampshire.[22] This configuration of a centralized computer or mainframe computer with remote "dumb terminals" remained popular throughout the 1950s and into the 60's. However, it was not until the 1960s that researchers started to investigate packet switching — a technology that allows chunks of data to be sent between different computers without first passing through a centralized mainframe. A four-node network emerged on December 5, 1969. This network soon became the ARPANET, which by 1981 would consist of 213 nodes.[23]

ARPANET's development centred around the Request for Comment process and on 7 April 1969, RFC 1 was published. This process is important because ARPANET would eventually merge with other networks to form the Internet, and many of the communication protocols that the Internet relies upon today were specified through the Request for Comment process. In September 1981, RFC 791 introduced the Internet Protocol version 4 (IPv4) and RFC 793 introduced the Transmission Control Protocol (TCP) — thus creating the TCP/IP protocol that much of the Internet relies upon today.

However, not all important developments were made through the Request for Comment process. Two popular link protocols for local area networks (LANs) also appeared in the 1970s. A patent for the token ring protocol was filed by Olof Soderblom on October 29, 1974, and a paper on the Ethernet protocol was published by Robert Metcalfe and David Boggs in the July 1976 issue of *Communications of the ACM*.[24] [25] The Ethernet protocol had been inspired by the ALOHAnet protocol which had been developed by electrical engineering researchers at the University of Hawaii.

Key concepts

A number of key concepts reoccur throughout the literature on modern telecommunication systems. Some of these concepts are discussed below.

Basic elements

A basic telecommunication system consists of three primary units that are always present in some form:

- A transmitter that takes information and converts it to a signal.
- A transmission medium, also called the "physical channel" that carries the signal. An example of this is the "free space channel".
- A receiver that takes the signal from the channel and converts it back into usable information.

For example, in a radio broadcasting station the station's large power amplifier is the transmitter; and the broadcasting antenna is the interface between the power amplifier and the "free space channel". The free space channel is the transmission medium; and the receiver's antenna is the interface between the free space channel and the receiver. Next, the radio receiver is the destination of the radio signal, and this is where it is converted from electricity to sound for people to listen to.

Sometimes, telecommunication systems are "duplex" (two-way systems) with a single box of electronics working as both a transmitter and a receiver, or a *transceiver*. For example, a cellular telephone is a transceiver.[26] The transmission electronics and the receiver electronics in a transceiver are actually quite independent of each other. This can be readily explained by the fact that radio transmitters contain power amplifiers that operate with electrical powers measured in the watts or kilowatts, but radio receivers deal with radio powers that are measured in the microwatts or nanowatts. Hence, transceivers have to be carefully designed and built to isolate their high-power circuitry and their low-power circuitry from each other.

Telecommunication over telephone lines is called point-to-point communication because it is between one transmitter and one receiver. Telecommunication through radio broadcasts is called broadcast communication because it is between one powerful transmitter and numerous low-power but sensitive radio receivers.[26]

Telecommunications in which multiple transmitters and multiple receivers have been designed to cooperate and to share the same physical channel are called multiplex systems.

Analog versus digital communications

Communications signals can be either by analog signals or digital signals. There are analog communication systems and digital communication systems. For an analog signal, the signal is varied continuously with respect to the information. In a digital signal, the information is encoded as a set of discrete values (for example, a set of ones and zeros). During the propagation and reception, the information contained in analog signals will inevitably be degraded by undesirable physical noise. (The output of a transmitter is noise-free for all practical purposes.) Commonly, the noise in a communication system can be expressed as adding or subtracting from the desirable signal in a completely random way. This form of noise is called ***"additive noise"***, with the understanding that the noise can be negative or positive at different instants of time. Noise that is not additive noise is a much more difficult situation to describe or analyze, and these other kinds of noise will be omitted here.

On the other hand, unless the *additive noise* disturbance exceeds a certain threshold, the information contained in digital signals will remain intact. Their resistance to noise represents a key advantage of digital signals over analog signals.[27]

Telecommunication networks

A communications network is a collection of transmitters, receivers, and communications channels that send messages to one another. Some digital communications networks contain one or more routers that work together to transmit information to the correct user. An analog communications network consists of one or more switches that establish a connection between two or more users. For both types of network, repeaters may be necessary to amplify or recreate the signal when it is being transmitted over long distances. This is to combat attenuation that can render the signal indistinguishable from the noise.[28]

Communication channels

The term "channel" has two different meanings. In one meaning, a channel is the physical medium that carries a signal between the transmitter and the receiver. Examples of this include the atmosphere for sound communications, glass optical fibers for some kinds of optical communications, coaxial cables for communications by way of the voltages and electric currents in them, and free space for communications using visible light, infrared waves, ultraviolet light, and radio waves. This last channel is called the "free space channel". The sending of radio waves from one place to another has nothing to do with the presence or absence of an atmosphere between the two. Radio waves travel through a perfect vacuum just as easily as they travel through air, fog, clouds, or any other kind of gas besides air.

The other meaning of the term "channel" in telecommunications is seen in the phrase communications channel, which is a subdivision of a transmission medium so that it can be used to send multiple streams of information simultaneously. For example, one radio station can broadcast radio waves into free space at frequencies in the neighborhood of 94.5 MHz (megahertz) while another radio station can simultaneously broadcast radio waves at frequencies in the neighborhood of 96.1 MHz. Each radio station would transmit radio waves over a frequency bandwidth of about 180 kHz (kilohertz), centered at frequencies such as the above, which are called the "carrier frequencies". Each station in this example is separated from its adjacent stations by 200 kHz, and the difference between 200 kHz and 180 kHz (20 kHz) is an engineering allowance for the imperfections in the communication system.

In the example above, the "free space channel" has been divided into communications channels according to frequencies, and each channel is assigned a separate frequency bandwidth in which to broadcast radio waves. This system of dividing the medium into channels according to frequency is called "frequency-division multiplexing" **(FDM)**.

Another way of dividing a communications medium into channels is to allocate each sender a recurring segment of time (a "time slot", for example, 20 milliseconds out of each second), and to allow each sender to send messages only within its own time slot. This method of dividing the medium into communication channels is called "time-division multiplexing" (**TDM**), and is used in optical fiber communication.[28] [29] Some radio communication systems use TDM within an allocated FDM channel. Hence, these systems use a hybrid of TDM and FDM.

Modulation

The shaping of a signal to convey information is known as modulation. Modulation can be used to represent a digital message as an analog waveform. This is commonly called "keying" − a term derived from the older use of Morse Code in telecommunications − and several keying techniques exist (these include phase-shift keying, frequency-shift keying, and amplitude-shift keying). The "Bluetooth" system, for example, uses phase-shift keying to exchange information between various devices.[30] [31] In addition, there are combinations of phase-shift keying and amplitude-shift keying which is called (in the jargon of the field) "quadrature amplitude modulation" (QAM) that are used in high-capacity digital radio communication systems.

Modulation can also be used to transmit the information of low-frequency analog signals at higher frequencies. This is helpful because low-frequency analog signals cannot be effectively transmitted over free space. Hence the information from a low-frequency analog signal must be impressed into a higher-frequency signal (known as the "carrier wave") before transmission. There are several different modulation schemes available to achieve this [two of the most basic being amplitude modulation (AM) and frequency modulation (FM)]. An example of this process is a disc jockey's voice being impressed into a 96 MHz carrier wave using frequency modulation (the voice would then be received on a radio as the channel "96 FM").[32] In addition, modulation has the advantage of being about to use frequency division multiplexing (FDM).

Society and telecommunication

Telecommunication has a significant social, cultural and economic impact on modern society. In 2008, estimates placed the telecommunication industry's revenue at $3.85 trillion or just under 3 percent of the gross world product (official exchange rate).[3] Several following sections discuss the impact of telecommunication on society.

Economic impact

Microeconomics

On the microeconomic scale, companies have used telecommunications to help build global business empires. This is self-evident in the case of online retailer Amazon.com but, according to academic Edward Lenert, even the conventional retailer Wal-Mart has benefited from better telecommunication infrastructure compared to its competitors.[33] In cities throughout the world, home owners use their telephones to order and arrange a variety of home services ranging from pizza deliveries to electricians. Even relatively poor communities have been noted to use telecommunication to their advantage. In Bangladesh's Narshingdi district, isolated villagers use cellular phones to speak directly to wholesalers and arrange a better price for their goods. In Côte d'Ivoire, coffee growers share mobile phones to follow hourly variations in coffee prices and sell at the best price.[34]

Macroeconomics

On the macroeconomic scale, Lars-Hendrik Röller and Leonard Waverman suggested a causal link between good telecommunication infrastructure and economic growth.[35] Few dispute the existence of a correlation although some argue it is wrong to view the relationship as causal.[36]

Because of the economic benefits of good telecommunication infrastructure, there is increasing worry about the inequitable access to telecommunication services amongst various countries of the world—this is known as the digital divide. A 2003 survey by the International Telecommunication Union (ITU) revealed that roughly a third of countries have fewer than one mobile subscription for every 20 people and one-third of countries have fewer than one land-line telephone subscription for every 20 people. In terms of Internet access, roughly half of all countries have fewer than one out of 20 people with Internet access. From this information, as well as educational data, the ITU was able to compile an index that measures the overall ability of citizens to access and use information and communication technologies.[37] Using this measure, Sweden, Denmark and Iceland received the highest ranking while the African countries Nigeria, Burkina Faso and Mali received the lowest.[38]

Social impact

Telecommunication has played a significant role in social relationships. Nevertheless devices like the telephone system were originally advertised with an emphasis on the practical dimensions of the device (such as the ability to conduct business or order home services) as opposed to the social dimensions. It was not until the late 1920s and 1930s that the social dimensions of the device became a prominent theme in telephone advertisements. New promotions started appealing to consumers' emotions, stressing the importance of social conversations and staying connected to family and friends.[39]

Since then the role that telecommunications has played in social relations has become increasingly important. In recent years, the popularity of social networking sites has increased dramatically. These sites allow users to communicate with each other as well as post photographs, events and profiles for others to see. The profiles can list a person's age, interests, sexual preference and relationship status. In this way, these sites can play important role in everything from organising social engagements to courtship.[40]

Prior to social networking sites, technologies like short message service(SMS) and the telephone also had a significant impact on social interactions. In 2000, market research group Ipsos MORI reported that 81% of 15 to 24 year-old SMS users in the United Kingdom had used the service to coordinate social arrangements and 42% to flirt.[41]

Other impacts

In cultural terms, telecommunication has increased the public's ability to access to music and film. With television, people can watch films they have not seen before in their own home without having to travel to the video store or cinema. With radio and the Internet, people can listen to music they have not heard before without having to travel to the music store.

Telecommunication has also transformed the way people receive their news. A survey led in 2006 by the non-profit Pew Internet and American Life Project found that when just over 3,000 people living in the United States were asked where they got their news "yesterday", more people said television or radio than newspapers. The results are summarised in the following table (the percentages add up to more than 100% because people were able to specify more than one source).[42]

Local TV	National TV	Radio	Local paper	Internet	National paper
59%	47%	44%	38%	23%	12%

Telecommunication has had an equally significant impact on advertising. TNS Media Intelligence reported that in 2007, 58% of advertising expenditure in the United States was spent on mediums that depend upon telecommunication.[43] The results are summarised in the following table.

	Internet	Radio	Cable TV	Syndicated TV	Spot TV	Network TV	Newspaper	Magazine	Outdoor	Total
Percent	7.6%	7.2%	12.1%	2.8%	11.3%	17.1%	18.9%	20.4%	2.7%	100%
Dollars	$11.31 billion	$10.69 billion	$18.02 billion	$4.17 billion	$16.82 billion	$25.42 billion	$28.22 billion	$30.33 billion	$4.02 billion	$149 billion

Telecommunication and government

Many countries have enacted legislation which conforms to the *International Telecommunication Regulations* established by the International Telecommunication Union (ITU), which is the "leading UN agency for information and communication technology issues."[44] In 1947, at the Atlantic City Conference, the ITU decided to "afford international protection to all frequencies registered in a new international frequency list and used in conformity with the Radio Regulation." According to the ITU's *Radio Regulations* adopted in Atlantic City, all frequencies referenced in the *International Frequency Registration Board*, examined by the board and registered on the *International Frequency List* "shall have the right to international protection from harmful interference."[45]

From a global perspective, there have been political debates and legislation regarding the management of telecommunication and broadcasting. The history of broadcasting discusses some debates in relation to balancing conventional communication such as printing and telecommunication such as radio broadcasting.[46] The onset of World War II brought on the first explosion of international broadcasting propaganda.[46] Countries, their governments, insurgents, terrorists, and militiamen have all used telecommunication and broadcasting techniques to promote propaganda.[46] [47] Patriotic propaganda for political movements and colonization started the mid 1930s. In 1936, the BBC did broadcast propaganda to the Arab World to partly counter similar broadcasts from Italy, which also had colonial interests in North Africa.[46]

Modern insurgents, such as those in the latest Iraq war, often use intimidating telephone calls, SMSs and the distribution of sophisticated videos of an attack on coalition troops within hours of the operation. "The Sunni insurgents even have their own television station, Al-Zawraa, which while banned by the Iraqi government, still broadcasts from Erbil, Iraqi Kurdistan, even as coalition pressure has forced it to switch satellite hosts several times." [47]

Modern telecommunication

Telephone

In an analog telephone network, the caller is connected to the person he wants to talk to by switches at various telephone exchanges. The switches form an electrical connection between the two users and the setting of these switches is determined electronically when the caller dials the number. Once the connection is made, the caller's voice is transformed to an electrical signal using a small microphone in the caller's handset. This electrical signal is then sent through the network to the user at the other end where it is transformed back into sound by a small speaker in that person's handset. There is a separate electrical connection that works in reverse, allowing the users to converse.[48] [49]

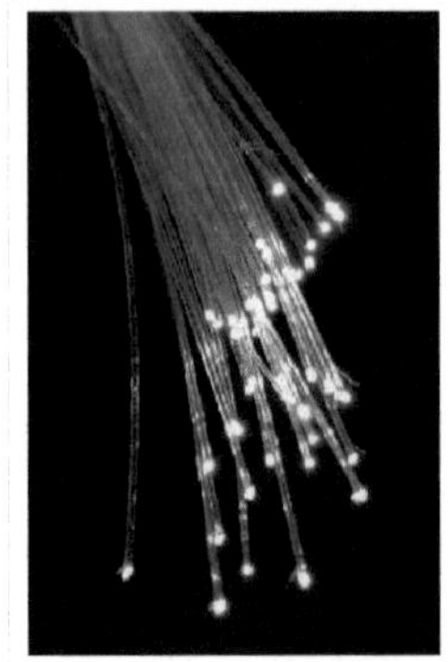

Optical fiber provides cheaper bandwidth for long distance communication

The fixed-line telephones in most residential homes are analog — that is, the speaker's voice directly determines the signal's voltage. Although short-distance calls may be handled from end-to-end as analog signals, increasingly telephone service providers are transparently converting the signals to digital for transmission before converting them back to analog for reception. The advantage of this is that digitized voice data can travel side-by-side with data from the Internet and can be perfectly reproduced in long distance communication (as opposed to analog signals that are inevitably impacted by noise).

Mobile phones have had a significant impact on telephone networks. Mobile phone subscriptions now outnumber fixed-line subscriptions in many markets. Sales of mobile phones in 2005 totalled 816.6 million with that figure being almost equally shared amongst the markets of Asia/Pacific (204 m), Western Europe (164 m), CEMEA (Central Europe, the Middle East and Africa) (153.5 m), North America (148 m) and Latin America (102 m).[50] In terms of new subscriptions over the five years from 1999, Africa has outpaced other markets with 58.2% growth.[51] Increasingly these phones are being serviced by systems where the voice content is transmitted digitally such as GSM or W-CDMA with many markets choosing to depreciate analog systems such as AMPS.[52]

There have also been dramatic changes in telephone communication behind the scenes. Starting with the operation of TAT-8 in 1988, the 1990s saw the widespread adoption of systems based on optic fibres. The benefit of communicating with optic fibers is that they offer a drastic increase in data capacity. TAT-8 itself was able to carry 10 times as many telephone calls as the last copper cable laid at that time and today's optic fibre cables are able to carry 25 times as many telephone calls as TAT-8.[53] This increase in data capacity is due to several factors: First, optic fibres are physically much smaller than competing technologies. Second, they do not suffer from crosstalk which means several hundred of them can be easily bundled together in a single cable.[54] Lastly, improvements in multiplexing have led to an exponential growth in the data capacity of a single fibre.[55] [56]

Assisting communication across many modern optic fibre networks is a protocol known as Asynchronous Transfer Mode (ATM). The ATM protocol allows for the side-by-side data transmission mentioned in the second paragraph. It is suitable for public telephone networks because it establishes a pathway for data through the network and associates a traffic contract with that pathway. The traffic contract is essentially an agreement between the client and the network about how the network is to handle the data; if the network cannot meet the conditions of the traffic contract it does not accept the connection. This is important because telephone calls can negotiate a contract so as to guarantee themselves a constant bit rate, something that will ensure a caller's voice is not delayed in parts or cut-off completely.[57] There are competitors to ATM, such as Multiprotocol Label Switching (MPLS), that perform a similar task and are expected to supplant ATM in the future.[58] [59]

Radio and television

In a broadcast system, the central high-powered broadcast tower transmits a high-frequency electromagnetic wave to numerous low-powered receivers. The high-frequency wave sent by the tower is modulated with a signal containing visual or audio information. The receiver is then tuned so as to pick up the high-frequency wave and a demodulator is used to retrieve the signal containing the visual or audio information. The broadcast signal can be either analog (signal is varied continuously with respect to the information) or digital (information is encoded as a set of discrete values).[26] [60]

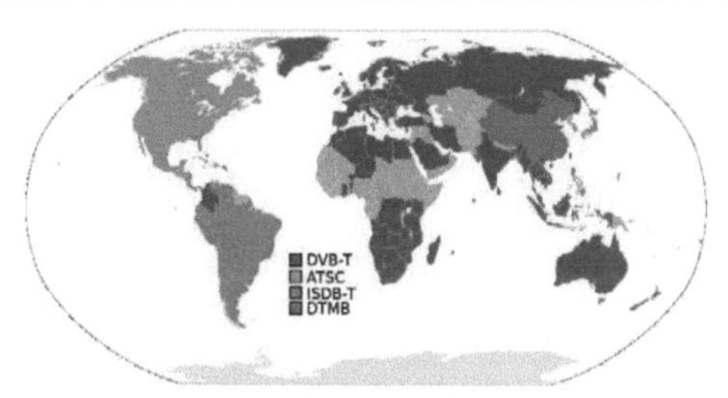

Digital television standards and their adoption worldwide.

The broadcast media industry is at a critical turning point in its development, with many countries moving from analog to digital broadcasts. This move is made possible by the production of cheaper, faster and more capable integrated circuits. The chief advantage of digital broadcasts is that they prevent a number of complaints common to traditional analog broadcasts. For television, this includes the elimination of problems such as snowy pictures, ghosting and other distortion. These occur because of the nature of analog transmission, which means that perturbations due to noise will be evident in the final output. Digital transmission overcomes this problem because digital signals are reduced to discrete values upon reception and hence small perturbations do not affect the final output. In a simplified example, if a binary message 1011 was transmitted with signal amplitudes [1.0 0.0 1.0 1.0] and received with signal amplitudes [0.9 0.2 1.1 0.9] it would still decode to the binary message 1011 — a perfect reproduction of what was sent. From this example, a problem with digital transmissions can also be seen in that if the noise is great enough it can significantly alter the decoded message. Using forward error correction a receiver can correct a handful of bit errors in the resulting message but too much noise will lead to incomprehensible output and hence a breakdown of the transmission.[61] [62]

In digital television broadcasting, there are three competing standards that are likely to be adopted worldwide. These are the ATSC, DVB and ISDB standards; the adoption of these standards thus far is presented in the captioned map. All three standards use MPEG-2 for video compression. ATSC uses Dolby Digital AC-3 for audio compression, ISDB uses Advanced Audio Coding (MPEG-2 Part 7) and DVB has no standard for audio compression but typically uses MPEG-1 Part 3 Layer 2.[63] [64] The choice of modulation also varies between the schemes. In digital audio broadcasting, standards are much more unified with practically all countries choosing to adopt the Digital Audio Broadcasting standard (also known as the Eureka 147 standard). The exception being the United States which has chosen to adopt HD Radio. HD Radio, unlike Eureka 147, is based upon a transmission method known as in-band on-channel transmission that allows digital information to "piggyback" on normal AM or FM analog transmissions.[65]

However, despite the pending switch to digital, analog television remains being transmitted in most countries. An exception is the United States that ended analog television transmission (by all but the very low-power TV stations) on 12 June 2009[66] after twice delaying the switchover deadline. For analog television, there are three standards in use for broadcasting color TV (see a map on adoption here). These are known as PAL (British designed), NTSC (North American designed), and SECAM (French designed). (It is important to understand that these are the ways from sending color TV, and they do not have anything to do with the standards for black & white TV, which also vary from country to country.) For analog radio, the switch to digital radio is made more difficult by the fact that analog receivers are sold at a small fraction of the price of digital receivers.[67] [68] The choice of modulation for analog radio is typically between amplitude modulation (**AM**) or frequency modulation (**FM**). To achieve stereo playback, an amplitude modulated subcarrier is used for stereo FM.

Internet

The Internet is a worldwide network of computers and computer networks that can communicate with each other using the Internet Protocol.[69] Any computer on the Internet has a unique IP address that can be used by other computers to route information to it. Hence, any computer on the Internet can send a message to any other computer using its IP address. These messages carry with them the originating computer's IP address allowing for two-way communication. The Internet is thus an exchange of messages between computers.[70]

It is estimated that the 51% of the information flowing through two-way telecommunications networks in the year 2000 were flowing through the Internet (most of the rest (42%) through the landline telephone). By the year 2007 the Internet clearly dominated and captured 97% of all the information in telecommunication networks (most of the rest (2%) through mobile phones).[1] As of 2008, an estimated 21.9% of the world population has access to the Internet with the highest access rates (measured as a percentage of the

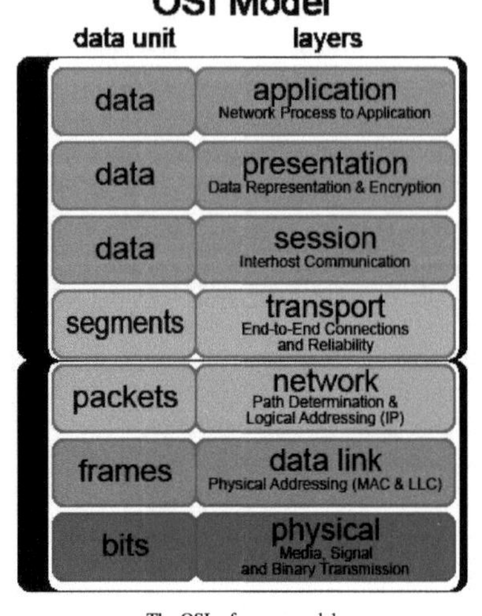

The OSI reference model

population) in North America (73.6%), Oceania/Australia (59.5%) and Europe (48.1%).[71] In terms of broadband access, Iceland (26.7%), South Korea (25.4%) and the Netherlands (25.3%) led the world.[72]

The Internet works in part because of protocols that govern how the computers and routers communicate with each other. The nature of computer network communication lends itself to a layered approach where individual protocols in the protocol stack run more-or-less independently of other protocols. This allows lower-level protocols to be customized for the network situation while not changing the way higher-level protocols operate. A practical example of why this is important is because it allows an Internet browser to run the same code regardless of whether the computer it is running on is connected to the Internet through an Ethernet or Wi-Fi connection. Protocols are often talked about in terms of their place in the OSI reference model (pictured on the right), which emerged in 1983 as the first step in an unsuccessful attempt to build a universally adopted networking protocol suite.[73]

For the Internet, the physical medium and data link protocol can vary several times as packets traverse the globe. This is because the Internet places no constraints on what physical medium or data link protocol is used. This leads to the adoption of media and protocols that best suit the local network situation. In practice, most intercontinental communication will use the Asynchronous Transfer Mode (ATM) protocol (or a modern equivalent) on top of optic fibre. This is because for most intercontinental communication the Internet shares the same infrastructure as the public switched telephone network.

At the network layer, things become standardized with the Internet Protocol (IP) being adopted for logical addressing. For the World Wide Web, these "IP addresses" are derived from the human readable form using the Domain Name System (e.g. 72.14.207.99 [74] is derived from www.google.com [75]). At the moment, the most widely used version of the Internet Protocol is version four but a move to version six is imminent.[76]

At the transport layer, most communication adopts either the Transmission Control Protocol (TCP) or the User Datagram Protocol (UDP). TCP is used when it is essential every message sent is received by the other computer whereas UDP is used when it is merely desirable. With TCP, packets are retransmitted if they are lost and placed in order before they are presented to higher layers. With UDP, packets are not ordered or retransmitted if lost. Both TCP and UDP packets carry port numbers with them to specify what application or process the packet should be handled by.[77] Because certain application-level protocols use certain ports, network administrators can manipulate traffic to suit particular requirements. Examples are to restrict Internet access by blocking the traffic destined for a particular port or to affect the performance of certain applications by assigning priority.

Above the transport layer, there are certain protocols that are sometimes used and loosely fit in the session and presentation layers, most notably the Secure Sockets Layer (SSL) and Transport Layer Security (TLS) protocols. These protocols ensure that the data transferred between two parties remains completely confidential and one or the other is in use when a padlock appears in the address bar of your web browser.[78] Finally, at the application layer, are many of the protocols Internet users would be familiar with such as HTTP (web browsing), POP3 (e-mail), FTP (file transfer), IRC (Internet chat), BitTorrent (file sharing) and OSCAR (instant messaging).

Voice over Internet Protocol (VoIP) allows data packets to be used for synchronous voice communications. The data packets are marked as voice type packets and can be prioritised by the network administrators so that the real-time, synchronous conversation is less subject to contention with other types of data traffic which can be delayed (i.e. file transfer or email) or buffered in advance (i.e. audio and video) without detriment. That prioritisation is fine when the network has sufficient capacity for all the VoIP calls taking place at the same time and the network is enabled for prioritisation i.e. a private corporate style network, but the Internet is not generally managed in this way and so there can be a big difference in the quality of VoIP calls over a private network and over the public Internet.[79]

Local area networks and wide area networks

Despite the growth of the Internet, the characteristics of local area networks ("LANs" − computer networks that do not extend beyond a few kilometers in size) remain distinct. This is because networks on this scale do not require all the features associated with larger networks and are often more cost-effective and efficient without them. When they are not connected with the Internet, they also have the advantages of privacy and security. However, purposefully lacking a direct connection to the Internet will not provide 100% protection of the LAN from hackers, military forces, or economic powers. These threats exist if there are any methods for connecting remotely to the LAN.

There are also independent wide area networks ("WANs" − private computer networks that can and do extend for thousands of kilometers.) Once again, some of their advantages include their privacy, security, and complete ignoring of any potential hackers − who cannot "touch" them. Of course, prime users of private LANs and WANs include armed forces and intelligence agencies that *must* keep their information completely secure and secret.

In the mid-1980s, several sets of communication protocols emerged to fill the gaps between the data-link layer and the application layer of the OSI reference model. These included Appletalk, IPX, and NetBIOS with the dominant protocol set during the early 1990s being IPX due to its popularity with MS-DOS users. TCP/IP existed at this point, but it was typically only used by large government and research facilities.[80]

As the Internet grew in popularity and a larger percentage of traffic became Internet-related, LANs and WANs gradually moved towards the TCP/IP protocols, and today networks mostly dedicated to TCP/IP traffic are common. The move to TCP/IP was helped by technologies such as DHCP that allowed TCP/IP clients to discover their own network address — a function that came standard with the AppleTalk/ IPX/ NetBIOS protocol sets.[81]

It is at the data-link layer, though, that most modern LANs diverge from the Internet. Whereas Asynchronous Transfer Mode (ATM) or Multiprotocol Label Switching (MPLS) are typical data-link protocols for larger networks such as WANs; Ethernet and Token Ring are typical data-link protocols for LANs. These protocols differ from the former protocols in that they are simpler (e.g. they omit features such as Quality of Service guarantees) and offer collision prevention. Both of these differences allow for more economical systems.[82] Despite the modest popularity of IBM token ring in the 1980s and 90's, virtually all LANs now use either wired or wireless Ethernets. At the physical layer, most wired Ethernet implementations use copper twisted-pair cables (including the common 10BASE-T networks). However, some early implementations used heavier coaxial cables and some recent implementations (especially high-speed ones) use optical fibers.[83] When optic fibers are used, the distinction must be made between multimode fibers and single-mode fiberes. Multimode fibers can be thought of as thicker optical fibers that are cheaper to manufacture devices for but that suffers from less usable bandwidth and worse attenuation – implying poorer long-distance performance.[84]

See also

- Active networks
- Busy Override
- Dual-tone multi-frequency signaling
- List of telecommunications billing companies
- Nanoscale networks
- Outline of telecommunication
- Push-button telephone
- Telecommunications Industry Association
- Telecoms resilience
- Wavelength-division multiplexing(WDM)
- Wired communication

References

[1] "The World's Technological Capacity to Store, Communicate, and Compute Information" (http://www.sciencemag.org/content/332/6025/ 60), Martin Hilbert and Priscila López (2011), Science, 332(6025), 60-65; free access to the study through here: martinhilbert.net/WorldInfoCapacity.html

[2] "video animation The Economist" (http://ideas.economist.com/video/giant-sifting-sound-0).

[3] Worldwide Telecommunications Industry Revenues (http://www.plunkettresearch.com/Telecommunications/ TelecommunicationsStatistics/tabid/96/Default.aspx), Internet Engineering Task Force, June 2010.

[4] *Telecommunication, tele-* and *communication*, New Oxford American Dictionary (2nd edition), 2005.

[5] Jean-Marie Dilhac, From tele-communicare to Telecommunications (http://www.ieee.org/portal/cms_docs_iportals/iportals/aboutus/ history_center/conferences/che2004/Dilhac.pdf), 2004.

[6] Lahanas, Michael, Ancient Greek Communication Methods (http://www.mlahanas.de/Greeks/Communication.htm), Mlahanas.de website. Retrieved July 14, 2009.

[7] David Ross, The Spanish Armada (http://www.britainexpress.com/History/tudor/armada.htm), Britain Express, October 2008.

[8] Les Télégraphes Chappe (http://chappe.ec-lyon.fr/), Cédrick Chatenet, l'Ecole Centrale de Lyon, 2003.

[9] CCIT/ITU-T 50 Years of Excellence (http://www.itu.int/itudoc/gs/promo/tsb/88192.pdf), International Telecommunication Union, 2006.

[10] The Electromagnetic Telegraph (http://www.du.edu/~jcalvert/tel/morse/morse.htm), J. B. Calvert, 19 May 2004.

[11] The Atlantic Cable (http://www.sil.si.edu/digitalcollections/hst/atlantic-cable/), Bern Dibner, Burndy Library Inc., 1959

[12] Sir Arthur C. Clarke. Voice Across the Sea (http://books.google.ca/books?id=L2UNAQAAIAAJ), Harper & Brothers, New York City, 1958.

[13] Brown, Travis (1994). *Historical first patents: the first United States patent for many everyday things* (http://books.google.com/ ?id=V-NUAAAAMAAJ&dq) (illustrated ed.). University of Michigan: Scarecrow Press. pp. 179. ISBN 978-0-8108-2898-8. .

[14] Connected Earth: The telephone (http://www.connected-earth.com/Galleries/Telecommunicationsage/Thetelephone/index.htm), BT, 2006.

[15] History of AT&T (http://www.att.com/history/milestones.html), AT&T, 2006.

[16] Dundee City Council. Biography: James Bowman Lindsay 1799 – 1862 (http://www.dundeecity.gov.uk/jbl/), Macdonald Black, Dundee City Council website. Retrieved September 9, 2010.

[17] Tesla Biography (http://www.teslasociety.com/biography.htm), Ljubo Vujovic, Tesla Memorial Society of New York, 1998.

[18] Tesla's Radio Controlled Boat (http://www.tfcbooks.com/teslafaq/q&a_025.htm), Twenty First Century Books, 2007.

[19] The Pioneers (http://www.mztv.com/newframe.asp?content=http://www.mztv.com/pioneers.html), MZTV Museum of Television, 2006.

[20] Philo Farnsworth (http://www.time.com/time/magazine/article/0,9171,990620,00.html), Neil Postman, *TIME Magazine*, 29 March 1999.

[21] Lotz, Amanda (2007). *The Television Will Be Revolutionized*. New York and London: New York University Press. pp. 3. ISBN 13:978-0-8147-5220-3.

[22] George Stlibetz (http://www.kerryr.net/pioneers/stibitz.htm), Kerry Redshaw, 1996.

[23] Hafner, Katie (1998). *Where Wizards Stay Up Late: The Origins Of The Internet*. Simon & Schuster. ISBN 0-684-83267-4.

[24] Data transmission system (http://patft1.uspto.gov/netacgi/nph-Parser?Sect1=PTO2&Sect2=HITOFF&p=1&u=/netahtml/PTO/search-bool.html&r=1&f=G&l=50&co1=AND&d=PTXT&s1=4293948.PN.&OS=PN/4293948&RS=PN/4293948), Olof Solderblom, PN 4,293,948, October 1974.

[25] Ethernet: Distributed Packet Switching for Local Computer Networks (http://www.acm.org/classics/apr96/), Robert M. Metcalfe and David R. Boggs, Communications of the ACM (pp 395–404, Vol. 19, No. 5), July 1976.

[26] Haykin, Simon (2001). *Communication Systems* (4th ed.). John Wiley & Sons. pp. 1–3. ISBN 0-471-17869-1.

[27] Ambardar, Ashok (1999). *Analog and Digital Signal Processing* (2nd ed.). Brooks/Cole Publishing Company. pp. 1–2. ISBN 0-534-95409-X.

[28] ATIS Telecom Glossary 2000 (http://www.atis.org/tg2k/), ATIS Committee T1A1 Performance and Signal Processing (approved by the American National Standards Institute), 28 February 2001.

[29] Yao, Colin (June 14, 2008). "Introduction to SONET (Synchronous Optical Networking)" (http://www.articlesbase.com/computers-articles/introduction-to-sonet-synchronous-optical-networking-fiber-optic-technologies-tutorial-series-449247.html). Articlesbase.com. .

[30] Haykin, pp 344–403.

[31] Bluetooth Specification Version 2.0 + EDR (http://www.bluetooth.org/foundry/adopters/document/Core_v2.0_EDR/en/1/Core_v2.0_EDR.zip) (p 27), Bluetooth, 2004.

[32] Haykin, pp 88–126.

[33] Lenert, Edward (10.1111/j.1460-2466.1998.tb02767.x). "A Communication Theory Perspective on Telecommunications Policy". *Journal of Communication* **48** (4): 3–23. doi:10.1111/j.1460-2466.1998.tb02767.x.

[34] Mireille Samaan (April 2003) (PDF). *The Effect of Income Inequality on Mobile Phone Penetration* (http://web.archive.org/web/20070214102055/http://dissertations.bc.edu/cgi/viewcontent.cgi?article=1016&context=ashonors). Boston University Honors thesis. Archived from the original (http://dissertations.bc.edu/cgi/viewcontent.cgi?article=1016&context=ashonors) on February 14, 2007. . Retrieved June 8, 2007.

[35] Röller, Lars-Hendrik; Leonard Waverman (2001). "Telecommunications Infrastructure and Economic Development: A Simultaneous Approach". *American Economic Review* **91** (4): 909–923. doi:10.1257/aer.91.4.909. ISSN 0002-8282.

[36] Riaz, Ali (10.1177/016344397019004004). "The role of telecommunications in economic growth: proposal for an alternative framework of analysis". *Media, Culture & Society* **19** (4): 557–583. doi:10.1177/016344397019004004.

[37] "Digital Access Index (DAI)" (http://www.itu.int/ITU-D/ict/dai/). itu.int. . Retrieved March 6, 2008.

[38] World Telecommunication Development Report 2003 (http://www.itu.int/ITU-D/ict/publications/wtdr_03/index.html), International Telecommunication Union, 2003.

[39] Fischer, Claude S.. "'Touch Someone': The Telephone Industry Discovers Sociability." Technology and Culture 29.1 (January 1988): 32–61. JSTOR. Web. 4 October 2009.

[40] "How do you know your love is real? Check Facebook" (http://www.cnn.com/2008/LIVING/personal/04/04/facebook.love/index.html). CNN. April 4, 2008. .

[41] I Just Text To Say I Love You (http://www.ipsos-mori.com/researchpublications/researcharchive/1575/I-Just-Text-To-Say-I-Love-You.aspx), Ipsos MORI, September 2005.

[42] "Online News: For many home broadband users, the internet is a primary news source" (http://www.pewinternet.org/pdfs/PIP_News.and.Broadband.pdf). Pew Internet Project. March 22, 2006. .

[43] "100 Leading National Advertisers" (http://adage.com/images/random/datacenter/2008/spendtrends08.pdf) (PDF). Advertising Age. June 23, 2008. . Retrieved June 21, 2009.

[44] International Telecommunication Union : About ITU (http://www.itu.int/net/about/index.aspx). ITU. Accessed 21 July 2009. (PDF (https://www.itu.int/osg/csd/wtpf/wtpf2009/documents/ITU_ITRs_88.pdf) of regulation)

[45] Codding, George A. Jr.. " *Jamming and the Protection of Frequency Assignments* (http://www.jstor.org/pss/2194872)". The American Journal of International Law, Vol. 49, No. 3 (Jul., 1955), Published by: American Society of International Law. pp. 384–388. Republished by JSTOR.org The American Journal of International Law". *Accessed 21 July 2009.*

[46] Wood, James & Science Museum (Great Britain) " *History of international broadcasting* (http://books.google.ca/books?id=WUO4U8L5N_cC&pg=PA3&lpg=PA3&dq="countries+use+telecommunications+for+propaganda&source=bl&ots=xZ23AbxMud&sig=elOIe1XUeivJ4fvrB5DPXDA6H54&hl=en&ei=CglmSpWjMIraNu3r9ZsB&sa=X&oi=book_result&ct=result&resnum=2)". IET 1994, Volume 1, p.2 of 258 ISBN 0-86341-302-1, ISBN 978-0-86341-302-5. Republished by Googlebooks. Accessed 21 July 2009.

[47] Garfield, Andrew. " *The U.S. Counter-propaganda Failure in Iraq* (http://www.meforum.org/1753/the-us-counter-propaganda-failure-in-iraq)", FALL 2007, The Middle East Quarterly, Volume XIV: Number 4, Accessed 21 July 2009.

[48] How Telephone Works (http://electronics.howstuffworks.com/telephone1.htm), HowStuffWorks.com, 2006.

[49] Telephone technology page (http://www.epanorama.net/links/telephone.html), ePanorama, 2006.

[50] Gartner Says Top Six Vendors Drive Worldwide Mobile Phone Sales to 21% Growth in 2005 (http://www.gartner.com/press_releases/asset_145891_11.html), Gartner Group, 28 February 2006.

[51] Africa Calling (http://www.spectrum.ieee.org/may06/3426), Victor and Irene Mbarika, IEEE Spectrum, May 2006.

[52] Ten Years of GSM in Australia (http://www.amta.org.au/default.asp?Page=142), Australia Telecommunications Association, 2003.

[53] Milestones in AT&T History (http://www.att.com/history/milestones.html), AT&T Knowledge Ventures, 2006.

[54] Optical fibre waveguide (http://www.cs.ucl.ac.uk/staff/S.Bhatti/D51-notes/node21.html), Saleem Bhatti, 1995.

[55] Fundamentals of DWDM Technology (http://www.cisco.com/univercd/cc/td/doc/product/mels/cm1500/dwdm/dwdm_ovr.pdf), CISCO Systems, 2006.

[56] Report: DWDM No Match for Sonet (http://www.lightreading.com/document.asp?doc_id=31358), Mary Jander, Light Reading, 2006.

[57] Stallings, William (2004). *Data and Computer Communications* (7th edition (intl) ed.). Pearson Prentice Hall. pp. 337–366. ISBN 0-13-183311-1.

[58] MPLS is the future, but ATM hangs on (http://www.networkworld.com/columnists/2002/0812edit.html), John Dix, Network World, 2002

[59] Lazar, Irwin (22 February 2011). "The WAN Road Ahead: Ethernet or Bust?" (http://www.telarus.com/industry/the-wan-road-ahead:-ethernet-or-bust.html). *Telecom Industry Updates*. . Retrieved 22 February 2011.

[60] How Radio Works (http://www.howstuffworks.com/radio.htm), HowStuffWorks.com, 2006.

[61] Digital Television in Australia (http://www.digitaltv.com.au/), Digital Television News Australia, 2001.

[62] Stallings, William (2004). *Data and Computer Communications* (7th edition (intl) ed.). Pearson Prentice Hall. ISBN 0-13-183311-1.

[63] HDV Technology Handbook (http://www.dynamix.ca/doc/HDVhandbook1.pdf), Sony, 2004.

[64] Audio (http://www.dvb.org/technology/standards_specifications/audio/), Digital Video Broadcasting Project, 2003.

[65] Status of DAB (USA) (http://www.worlddab.org/cstatus.aspx), World DAB Forum, March 2005.

[66] Brian Stelter (June 13, 2009). "Changeover to Digital TV Off to a Smooth Start" (http://www.nytimes.com/2009/06/14/business/media/14digital.html?_r=2&hp). *New York Times*. .

[67] GE 72664 Portable AM/FM Radio (http://www.amazon.com/dp/B00000J060), Amazon.com, June 2006.

[68] DAB Products (http://www.worlddab.org/dabprod.aspx), World DAB Forum, 2006.

[69] Robert E. Kahn and Vinton G. Cerf, What Is The Internet (And What Makes It Work) (http://www.cnri.reston.va.us/what_is_internet.html), December 1999. (specifically see footnote xv)

[70] How Internet Infrastructure Works (http://computer.howstuffworks.com/internet-infrastructure.htm), HowStuffWorks.com, 2007.

[71] World Internet Users and Population Stats (http://www.internetworldstats.com/stats.htm), internetworldstats.com, 19 March 2007.

[72] OECD Broadband Statistics (http://www.oecd.org/document/39/0,2340,en_2649_34225_36459431_1_1_1_1,00.html), Organisation for Economic Co-operation and Development, December 2005.

[73] History of the OSI Reference Model (http://www.tcpipguide.com/free/t_HistoryoftheOSIReferenceModel.htm), The TCP/IP Guide v3.0, Charles M. Kozierok, 2005.

[74] http://72.14.207.99/

[75] http://www.google.com/

[76] Introduction to IPv6 (http://www.microsoft.com/technet/itsolutions/network/ipv6/introipv6.mspx), Microsoft Corporation, February 2006.

[77] Stallings, pp 683–702.

[78] T. Dierks and C. Allen, The TLS Protocol Version 1.0, RFC 2246, 1999.

[79] Voice over Internet Protocol (VoIP) and Internet Telephony http://www.telecomsadvice.org.uk/infosheets/voip_voice_over_internet_protocol_and_internet_telephony.htm

[80] Martin, Michael (2000). *Understanding the Network* (The Networker's Guide to AppleTalk, IPX, and NetBIOS (http://www.informit.com/content/images/0735709777/samplechapter/0735709777.pdf)), SAMS Publishing, ISBN 0-7357-0977-7.

[81] Ralph Droms, Resources for DHCP (http://www.dhcp.org/), November 2003.

[82] Stallings, pp 500–526.

[83] Stallings, pp 514–516.

[84] Fiber Optic Cable Tutorial (http://www.arcelect.com/fibercable.htm), Arc Electronics. Retrieved June, 2007.

Further reading

- OECD, *Universal Service and Rate Restructuring in Telecommunications* (http://books.google.com/books?id=WpmzcqmgMbAC&dq=universal+service+and+rate+restructuring+in+telecommunications&printsec=frontcover&source=bl&ots=S2USGNAune&sig=Alh7pDRwI3Rk4iYVYuMq9rZlIZc&hl=en&sa=X&oi=book_result&resnum=1&ct=result#PPP1,M1), Organisation for Economic Co-operation and Development (OECD) Publishing, 1991. ISBN 92-64-13497-2

- Wheen, Andrew. DOT-DASH TO DOT.COM: How Modern Telecommunications Evolved from the Telegraph to the Internet (Springer, 2011)

External links

- ATIS Telecom Glossary (http://www.atis.org/tg2k/)
- Communications Engineering Tutorials (http://www.complextoreal.com/tutorial.htm)
- Federal Communications Commission (http://www.fcc.gov/)
- Unified Communications (http://www.radvision.com/Unified-Communications/)
- IEEE Communications Society (http://www.comsoc.org/)
- International Telecommunication Union (http://www.itu.int/home/)
- Ericsson's Understanding Telecommunications (http://web.archive.org/web/20040413074912/www.ericsson.com/support/telecom/index.shtml) at archive.org (Ericsson removed the book from their site in September 2005)
- VoIP, Voice over Internet Protocol and Internet telephone calls (http://www.telecomsadvice.org.uk/infosheets/voip_voice_over_internet_protocol_and_internet_telephony.htm)

Coverage_map

Coverage maps are designed to indicate the service areas of radiocommunication transmitting stations. Typically these may be produced for radio or television stations, for mobile telephone networks and for satellite networks. Such maps are alternatively known as propagation maps. For satellite networks, a coverage map is often known as a footprint.

Definition of coverage

Typically a coverage map will indicate the area within which the user can expect to obtain good reception of the service in question using standard equipment under normal operating conditions. Additionally, the map may also separately denote supplementary service areas where good reception may be obtained but other stations may be stronger, or where reception may variable but the service may still be usable.

Technical details

The field strength that the marked service boundary on a coverage map represents will be defined by whoever produces the map, but typical examples are as follows:

VHF(FM) / Band II

For VHF(FM) / Band II, the BBC defines the service area boundary as corresponding to an average field strength of 54 dB (relative to 1 µV/m) at a height of 10 m above ground level.

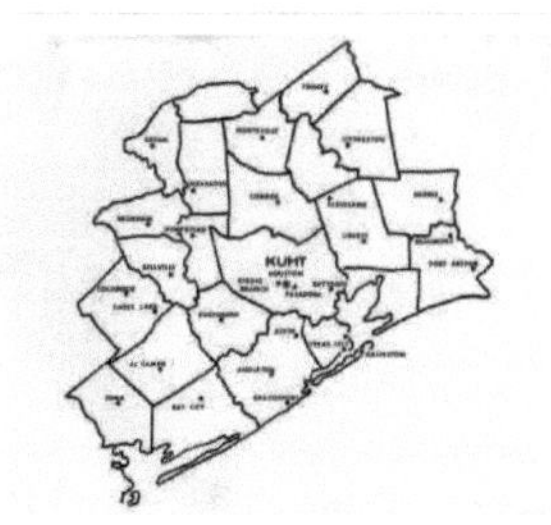

A past coverage map of Alvin, Texas

MF / Mediumwave

For MF / Mediumwave, the BBC defines the daytime service area boundary as a minimum field strength of 2 mV/m. At night, the service area of mediumwave services can be drastically reduced by co-channel interference from distant stations.

Limitations

Often coverage maps show general coverage for large regions and therefore any boundary indicated should not be interpreted as a rigid limit. The quality of reception can be very different at places only short distances apart, and this phenomenon is more apparent as the transmission frequency increases. Inevitably small pockets of poor reception may exist within the main service area that cannot be shown on the map due to scale issues. Conversely, the use of sensitive equipment, high gain antennas, or simply being located on high ground can yield good signal strengths well outside the indicated area. The significance of local geographical conditions cannot be over emphasised and this was underlined by an experiment [1] which revealed the signal reception conditions around a typical house. The site did not have the critical "line-of-sight propagation" to the transmitter. Average signal levels, taken at the same height, varied by up to 6dB, and for individual frequencies by up to 14dB. In RF reception terms these figures are huge differences.

Although carriers and broadcasters attempt to design their networks to eliminate dead zones, no network is perfect, so coverage breaks within the general coverage areas are still possible.

Often companies will construct low power satellite stations to fill in bad reception areas that become apparent once the high power transmitter's coverage map has identified where the network is deficient.

External links

- The Transmission Gallery: Index of UK TV coverage maps [2]
- TV Fool: Index of US TV coverage maps [3]

References

[1] http://www.aerialsandtv.com/loftaerials.html#BlackArtOfRF
[2] http://tx.mb21.co.uk/mapsys/anatv/
[3] http://www.tvfool.com/index.php?option=com_content&task=view&id=15

Orography

Orography (from the Greek ὄρος, hill, γραφία, to write) is the study of the formation and relief of mountains,[1] and can more broadly include hills, and any part of a region's elevated terrain.[2] Orography (also known as *oreography*, *orology* or *oreology*) falls within the broader discipline of geomorphology.

Uses

Orography has a major impact on global climate, for instance the orography of East Africa substantially determines the strength of the Indian monsoon.[3] In geoscientific models, such as general circulation models, orography defines the lower boundary of the model over land.

When a river's tributaries or settlements by the river are listed in 'orographic sequence', they are in order from the highest (nearest the source of the river) to the lowest or mainstem (nearest the mouth). This method of listing tributaries is similar to the Strahler Stream Order, where the headwater tributaries are listed as category = 1.

Precipitation

Orographic precipitation, also known as relief precipitation, is precipitation generated by a forced upward movement of air upon encountering a physiographic upland (see anabatic wind). This lifting can be caused by two mechanisms:

1. The upward deflection of large scale horizontal flow by the orography.
2. The anabatic or upward vertical propagation of moist air up an orographic slope caused by daytime heating of the mountain barrier surface.

Upon ascent, the air that is being lifted will expand and cool. This adiabatic cooling of a rising moist air parcel may lower its temperature to its dew point, thus allowing for condensation of the water vapor contained within it, and hence the formation of a cloud. If enough water vapor condenses into cloud droplets, these droplets may become large enough to fall to the ground as precipitation. In parts of the world subjected to relatively consistent winds (for example the trade winds), a wetter climate prevails on the windward side of a mountain than on the leeward (downwind) side as moisture is removed by orographic precipitation. Drier air (see katabatic wind) is left on the descending, generally warming, leeward side where a rain shadow is formed.

Terrain induced precipitation is a major factor for meteorologists as they forecast the local weather. Orography can play a major role in the type, amount, intensity and duration of precipitation events. Researchers have discovered that barrier width, slope steepness and updraft speed are major contributors for the optimal amount and intensity of orographic precipitation. Computer model simulations for these factors showed that narrow barriers and steeper slopes produced stronger updraft speeds which, in turn, enhanced orographic precipitation.

Orographic precipitation is well known on oceanic islands, such as the Hawaiian Islands or New Zealand, where much of the rainfall received on an island is on the windward side, and the leeward side tends to be quite dry, almost desert-like, by comparison. This phenomenon results in substantial local gradients of average rainfall, with coastal areas receiving on the order of 20 to 30 inches (**unknown operator: u'strong'** to **unknown operator: u'strong'** mm) per year, and interior uplands receiving over 100 inches (**unknown operator: u'strong'** mm) per year. Leeward coastal areas are especially dry—less than 20 in (**unknown operator: u'strong'** mm) per year at Waikiki—and the tops of moderately high uplands are especially wet—about 475 in (**unknown operator: u'strong'** mm) per year at Wai'ale'ale on Kaua'i.

Another well known area for orographic precipitation is the Pennines in the north of England where the west side of the Pennines receives more rain than the east because the clouds (generally arriving from the west) are forced up and over the hills and cause the rain to fall preferentially on the western slopes. This is particularly noticeable between Manchester (West) and Leeds (East) where Leeds receives less rain due to a rain shadow of 12 miles from the

Pennines.

See also

- Coverage (telecommunication)
- Orographic lift

References

[1] 11th Edition of Encyclopaedia Britannica (1911) (http://www.1911encyclopedia.org/Orography)

[2] Orography (http://amsglossary.allenpress.com/glossary/search?p=1&query=Orography&submit=Search) (from the American Meteorological Society website)

[3] Srinivasan, J., Nanjundiah, Ravi S. and Chakraborty, Arindam (2005) Impact of Orography on the Simulation of Monsoon Climate in a General Circulation Model (http://hdl.handle.net/2005/76) *Indian Institute of Science*

External links

- Map of the Orography of Europe (http://www.euratlas.com/Atlasphys/Orography.htm) from Euratlas.com

Multi-storey_car_park

A **multi-storey car-park** (also called a **parking garage**, **parking structure**, **parking ramp**, **parkade**, **parking building** or **parking deck**) is a building designed for car parking and where there are a number of floors or levels on which parking takes place. It is essentially a stacked car park.

Busch Stadium exit ramp

Nomenclature

The term *multi-storey car park* is used in the United Kingdom, Hong Kong and many Commonwealth of Nations countries. In the western United States, the term *parking structure* is used, especially when it is necessary to distinguish such a structure from the "garage" in a house. In some places in North America, "parking garage" refers only to an indoor, often underground, structure – outdoor multi-level parking facilities are referred to by a number of regional terms:

- *Parking garage* is used in the Western United States, Eastern Canada and by civil engineers;
- *Parking deck* is used in the Southeast;
- *Parking ramp* is used in the upper Midwest, especially Minnesota and Wisconsin, and has been observed as far east as Buffalo, New York.
- *Parkade* is used in Canada and South Africa
- *Parking building* is used in New Zealand.

Architects and civil engineers in the USA are likely to call it a **parking structure**, since their work is all about structures, and that term is the vernacular in some of the western United States. When attached to a high-rise of another use, it is sometimes called a **parking podium**. United States building codes use the term **open parking**

structure to refer to a structure designed for car storage (not repair) that has enough openings in the walls that it does not need mechanical ventilation or fire sprinklers, as opposed to a "parking garage" that requires mechanical ventilation or sprinklers but does not require openings in the walls. The openings provide fresh air flow to disperse either car exhaust or fumes from a fire should one break out within the structure.

First multi-storey car park

The earliest known multi-storey car park was built in 1918 for the Hotel La Salle at 215 West Washington Street in the West Loop area of downtown Chicago, Illinois. It was designed by Holabird and Roche.[1] The Hotel La Salle was demolished in 1976, but the parking structure remained because it had been designated as preliminary landmark status[2] and the structure was several blocks from the hotel. It was demolished in 2005 after failing to receive landmark status from the city of Chicago.[3] Jupiter Realty Corp. of Chicago is constructing a 49-storey apartment tower in its place,[4] with construction underway as of March 2008.

An alternative claim has emerged from Glasgow, Scotland, for a building that was built between 1906 and 1912.[5]

In the 1920s an English cartoonist imagined a hotel for cars; he drew a multi-storey car park.[6]

Design

The inside of a multi-storey car park

Movement of vehicles between floors can be effected by:

Insulation of a sectional garage door

- interior ramps - the most common type
- exterior ramps - which may take the form of a circular ramp (colloquially known as a 'whirley-gig' in America)
- vehicle lifts - the least common
- automated robot systems - combination of ramp and elevator

Where the car park is built on sloping land, it may be split-level or have sloped parking.

Many car parks are independent buildings dedicated exclusively to that use. The design loads for car parks are often less than the office building they serve (50 psf versus 80 psf), leading to long floor spans of 55–60 feet that permit cars to park in rows without supporting columns in between. The most common structural systems in the United States for these structures are either prestressed concrete concrete double-tee floor systems or post-tensioned cast-in-place concrete floor systems.

In recent times, car parks built to serve residential and some business properties have been built as part of a larger building, often underground as part of the basement, such as at the Atlantic Station redevelopment in Atlanta. This saves land for other uses (as opposed to a parking lot), is cheaper and more practical in most cases than a separate structure, and is hidden from view. It protects customers and their cars from weather such as rain, snow, or hot summer sunshine that raises a vehicle's interior temperature to extremely high levels. Underground parking of only two levels was considered an innovative concept in 1964, when developer Louis Lesser developed a two-level underground parking structure under six 10-storey high-rise residential halls at California State University, Los Angeles, which lacked space for horizontal expansion in the 176-acre (**unknown operator: u'strong'** km^2) university. The simple two-level parking structure was considered unusual enough in 1964 that a separate newspaper section entitled "Parking Underground" described the garage as an innovative "concept" and as "subterranean spaces".[7] [8] In Toronto, a 2,400 space parking lot below Nathan Phillips Square is one of the world's largest.

Car parks which serve shopping centres can be built adjacent to the centre for easier access at each floor between shops and parking. One example is Mall of America in Bloomington, Minnesota, USA, which has two large car parks attached to the building, at the eastern and western ends. A common position for car parks within shopping centres in the UK is on the roof, around the various utility systems, enabling customers to take lifts straight down into the centre. Examples of such are The Oracle in Reading and Festival Place in Basingstoke.

Motorcycle parking inside a multi-storey car park

Structural integrity

Parking structures are subjected to the heavy and shifting loads of moving vehicles, and must bear the associated physical stresses. Expansion joints are used between sections not only for thermal expansion but to accommodate the flexing of the structure's sections due to vehicle traffic. Seismic retrofits can be applied where earthquakes are an issue.

Some parking structures have partly collapsed, either during construction or years later. In July 2009 a fourth-floor section failed at the Centergy building in midtown Atlanta, pancaking down and destroying more than 30 vehicles but injuring no-one. In December 2007, a car crashed into the wall of the deck at the SouthPark Mall in Charlotte, North Carolina, weakening it and causing a small collapse which destroyed two cars below. On the same day, one under construction in Jacksonville, Florida collapsed as concrete was being poured on the sixth floor.[9] In November 2008, the sudden collapse of the middle level of a deck in Montreal was preceded by warning signs some weeks before, including cracks and water leaks.[10] Parking structures are generally not subject to building inspections after being checked for their initial occupancy permit.

Architectural value

These structures are not usually known for their architectural value. As *Architectural Record* has noted, "In the Pantheon of Building Types, the parking garage lurks somewhere in the vicinity of prisons and toll plazas."[11] *The New York Times* has labeled parking structures as "the grim afterthought of American design".[12]

A handful of parking garages have has received considerable praise for their design, such as 1111 Lincoln Road, in the South Beach section of Miami Beach, Florida and designed by the internationally-known Swiss architectural firm of Herzog & de Meuron.[11] [12] . Another is the Santa Monica Place Public Parking Structures

The 1111 Lincoln Road parking garage

at Frank Gehry's Santa Monica Place Mall, by the architectural firm Brooks + Scarpa.

Santa Monica Place Public Parking Structures, Santa Monica, California

Automated parking

The first mention of an automated garage was in a 1931 Popular Mechanics article which featured an underground garage where the car was taken to a parking area by a conveyor then an elevator to shuttles mounted on rails [13] The first Kent Automatic Garages were already in service.

Automatic multi-storey car parks provide lower building cost per parking slot, as they typically require less building volume and less ground area than a conventional facility with the same capacity. However, the cost of the mechanical equipment needed to transport the cars needs to be added to the building cost to determine the total cost. Other costs are usually lower too, for example there is no need for an energy-intensive ventilating system, since cars are not driven inside and human cashiers or security personnel may not be needed.

Automatic underground car storage in Thessaloniki, Greece

Automated car parks rely on similar technology to that used for mechanical handling and document retrieval. The driver leaves the car in an entrance module, and it is then transported to a parking slot by a robot trolley. For the driver, the process of parking is reduced to leaving the car inside an entrance module.

At peak periods a wait may occur before entering or leaving because loading passengers and luggage occurs at the entrance and exit rather than at the parking stall. This loading blocks the entrance or exit from being available to others. Whether the retrieval of vehicles is faster in an automatic car park or a conventional car park depends on the layout and number of exits.

Modular car park

Parking demand often grows quickly, significantly and unexpectedly. Modular steel car parks could be the proper solution if the surface area available is not sufficient and can be expanded upwards, or whenever it is not feasible to build up a multi-storey parking. The development concept of traditional build modular car parks is made by the modular assembling method of vertical and horizontal elements (such as columns and beams) with a ceiling made of concrete and tarmac: more modular units can build a parking in different sizes and shape. The solution makes possible to develop a parking structure even in case of particular conditions or constraints, such as archaeological sites or city centres, because it allows:

- To virtually double the parking surface without leaving any footprint on the ground, as no settlement for excavations or traditional foundations is needed;
- To double the parking surface by means of a light steel single-deck car park system.
- Prefab modular components of the system make each project versatile and suitable both to large and small sized areas.

These parking structures are generally demountable and can be relocated so to avoid to make the choice of converting a surface to parking area irrevocably. They are conceived as temporary parking facilities for temporary parking demand needs, whenever the parking demand can be managed dynamically and easily integrated into the planning of urban infrastructures. A number of parking decks have been demounted after a few years, to make room to the development of permanent, multi-storey parking structures, and relocated to respond to local parking demand.

See also

- Autostadt
- Auto Stacker
- Automatic parking
- Car Parking System
- Car condo
- Parking guidance and information
- Trinity Square Gateshead

References

[1] A Brief History of Parking: The Life and After-life of Paving the Planet (http://www.asphaltnation.com/Articles/parking.html)

[2] Preservation Online: Storey of the Week Archives: Car Culture (http://www.nationaltrust.org/magazine/archives/arch_story/102105. htm)

[3] Preservation Chicago (http://www.preservationchicago.org/risk/lasalle.html)

[4] Chicago Business News, Analysis & Articles | Developer planning 2nd apartment tower downtown | Crain's (http://chicagobusiness.com/ cgi-bin/news.pl?id=24357)

[5] "Consultation over historic garage" (http://news.bbc.co.uk/1/hi/scotland/glasgow_and_west/7459839.stm). BBC News. 17 June 2008. . Retrieved 13 May 2010.

[6] Daniele Luttazzi *Lepidezze postribolari* (2007, Feltrinelli, p.275) (**Italian**)

[7] Los Angeles Times, 25 June 1963, "High Rise Developer Defends Loss of View to Convenience"

[8] Los Angeles Times, 15 March 1964, Tom Cameron, "$118 Million Going Into Expansion at L.A. State"

[9] "North Carolina Parking Deck Collapse Kills 1" (http://www.foxnews.com/story/0,2933,315587,00.html). Fox News. 6 December 2007. .

[10] "Man dies after Montreal parking garage roof collapses" (http://www.cbc.ca/canada/montreal/story/2008/11/26/mtl-bldg.html). *CBC News*. 26 November 2008. .

[11] Beth Broome (June 2010). "1111 Lincoln Road" (http://archrecord.construction.com/projects/portfolio/archives/ 10061111Lincoln_Road-1.asp). *Architectural Record*. .

[12] Michael Barbaro (January 24, 2011). "A Miami Beach Event Space. Parking Space, Too." (http://www.nytimes.com/2011/01/24/us/
24garage.html). *The New York Times*: p. A1. .

[13] "Underground Garage To End Parking Problem", February 1931, Popular Mechanics (http://books.google.com/
books?id=u-IDAAAAMBAJ&pg=PA180&dq=Popular+Mechanics+1931+curtiss&hl=en&ei=3bTyTKD7AZPfnQeEr635Cg&sa=X&
oi=book_result&ct=result&resnum=3&ved=0CDAQ6AEwAg#v=onepage&q&f=true)

External links

- "Automatic invisible compact parking. Patent RU 97419" (http://artamonoff.info/parking).
- "A brief history of parking" (http://www.asphaltnation.com/Articles/parking.html). Retrieved 22 February 2007.
- Mary Beth Klatt (21). "Car culture" (http://web.archive.org/web/20070211074000/http://www.nationaltrust.org/magazine/archives/arch_story/102105.htm). *Preservation Online*. National Trust for Historic Preservation. pp. 1. Archived from the original (http://www.nationaltrust.org/magazine/archives/arch_story/102105.htm) on 11 February 2007. Retrieved 22 February 2007.
- "Robotic Parking Garage: No Tip Necessary " (http://www.popularmechanics.com/technology/gadgets/4213198)
- "Automated parking" (http://www.territorioscuola.com/youtube/index.php?key=Automatisches Parkhaus).

Base_station

The term **base station** can be used in the context of land surveying and wireless communications.

Land surveying

In the context of external land surveying, a base station is a GPS receiver at an accurately-known fixed location which is used to derive correction information for nearby portable GPS receivers. This correction data allows propagation and other effects to be corrected out of the position data obtained by the mobile stations, which gives greatly increased location precision and accuracy over the results obtained by uncorrected GPS receivers.

A 1980s consumer-grade Citizens' band radio (CB) base station

Computer networking

In the area of wireless computer networking, a base station is a radio receiver/transmitter that serves as the hub of the local wireless network, and may also be the gateway between a wired network and the wireless network. It typically consists of a low-power transmitter and wireless router.

Wireless communications

In radio communications, a base station is a wireless communications station installed at a fixed location and used to communicate as part of either:

- a push-to-talk two-way radio system, or;
- a wireless telephone system such as cellular CDMA or GSM cell site.
- Terrestrial Trunked Radio

Two-way radio

Professional

In professional two-way radio systems, a base station is used to maintain contact with a dispatch fleet of hand-held or mobile radios, and/or to activate one-way paging receivers. The base station is one end of a communications link. The other end is a movable vehicle-mounted radio or walkie-talkie.[1] Examples of base station uses in two-way radio include the dispatch of tow trucks and taxicabs.

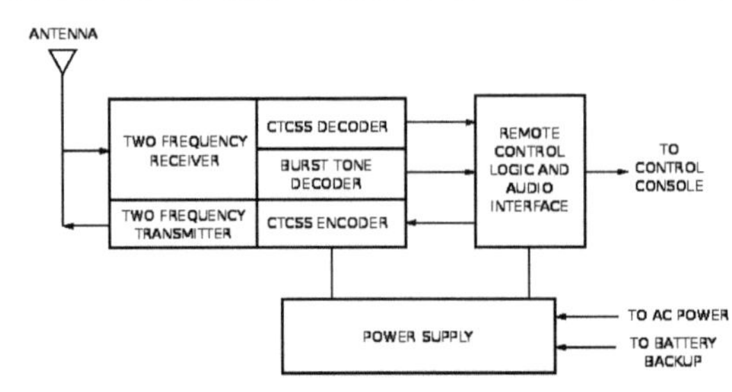

Basic base station elements used in a remote-controlled installation. Selective calling options such as CTCSS are optional.

Professional base station radios are often one channel. In lightly used base stations, a multi-channel unit may be employed.[2] In heavily used systems, the capability for additional channels, where needed, is accomplished by installing an additional base station for each channel. Each base station appears as a single channel on the dispatch center control console. In a properly designed dispatch center with several staff members, this allows each dispatcher to communicate simultaneously, independently of one another, on a different channel as necessary. For example, a taxi company dispatch center may have one base station on a high-rise building in Boston and another on a different channel in Providence. Each taxi dispatcher could communicate with taxis in either Boston or Providence by selecting the respective base station on his or her console.[3]

In dispatching centers it is common for eight or more radio base stations to be connected to a single dispatching console. Dispatching personnel can tell which channel a message is being received on by a combination of local protocol, unit identifiers, volume settings, and busy indicator lights. A typical console has two speakers identified as *select* and *unselect*. Audio from a primary selected channel is routed to the select speaker and to a headset. Each channel has a busy light which flashes when someone talks on the associated channel.[4]

Base stations can be local controlled or remote controlled. Local controlled base stations are operated by front panel controls on the base station cabinet. Remote control base stations can be operated over tone- or DC-remote circuits. The dispatch point console and remote base station are connected by leased private line telephone circuits, (sometimes called *RTO circuits*), a DS-1, or radio links.[5] The consoles multiplex transmit commands onto remote control circuits. Some system configurations require duplex, or four wire, audio paths from the base station to the console. Others require only a two-wire or half duplex link.[6]

Interference could be defined as receiving any signal other than from a radio in your own system. To avoid interference from users on the same channel, or interference from nearby strong signals on another channel, professional base stations use a combination of:[7] [8]

- minimum receiver specifications and filtering.[9] [10] [11]
- analysis of other frequencies in use nearby.
- in the US, coordination of shared frequencies by coordinating agencies.[12]
- locating equipment so that terrain blocks interfering signals.
- use of directional antennas to reduce unwanted signals.

Base stations are sometimes called *control* or *fixed* stations in US Federal Communications Commission licensing. These terms are defined in regulations inside Part 90 of the commissions regulations. In US licensing jargon, types of base stations include:

- A **fixed** station is a base station used in a system intended only to communicate with other base stations. A *fixed* station can also be radio link used to operate a distant base station by remote control. (No mobile or hand-held radios are involved in the system.)
- A **control** station is a base station used in a system with a repeater where the base station is used to communicate through the repeater.
- A **temporary base** is a base station used in one location for less than a year.
- A **repeater** is a type of base station that extends the range of hand-held and mobile radios.

The diagram shows a band-pass filter used to reduce the base station receiver's exposure to unwanted signals. It also reduces the transmission of undesired signals. The isolator is a one-way device which reduces the ease of signals from nearby transmitters going up the antenna line and into the base station transmitter. This prevents the unwanted mixing of signals inside the base station transmitter which can generate interference.

Amateur and hobbyist use

In amateur radio, a base station also communicates with mobile rigs but for hobby or family communications. Amateur systems sometimes serve as dispatch radio systems during disasters, search and rescue mobilizations, or other emergencies.

An Australian UHF CB base station is another example of part of a system used for hobby or family communications.

Wireless telephone

Wireless telephone differ from two-way radios in that:

- wireless telephones are circuit switched: the communications paths are set up by dialing at the start of a call and the path remains in place until one of the callers hangs up.
- wireless telephones communicate with other telephones usually over the public switched telephone network.

A wireless telephone base station communicates with a mobile or hand-held phone. For example, in a wireless telephone system, the signals from one or more mobile telephones in an area are received at a nearby base station, which then connects the call to the land-line network. Other equipment is involved depending on the system architecture. Mobile telephone provider networks, such as European GSM networks, may involve carrier, microwave radio, and switching facilities to connect the call. In the case of a portable phone such as a US cordless phone, the connection is directly connected to a wired land line.

Emissions issues

While low levels of radio-frequency power are usually considered to have negligible effects on health, national and local regulations restrict the design of base stations to limit exposure to electromagnetic fields. Technical measures to limit exposure include restricting the radio frequency power emitted by the station, elevating the antenna above ground level, changes to the antenna pattern, and barriers to foot or road traffic. For typical base stations, significant electromagnetic energy is only emitted at the antenna, not along the length of the antenna tower.[13]

Because mobile phones and their base stations are two-way radios, they produce radio-frequency (RF) radiation in order to communicate, exposing people near them to RF radiation giving concerns about mobile phone radiation and health. Hand-held mobile telephones are relatively low power so the RF radiation exposures from them are generally low.

A cell tower near Thicketty, South Carolina.

The World Health Organization has concluded that "there is no convincing scientific evidence that the weak RF signals from base stations and wireless networks cause adverse health effects."[14]

The consensus of the scientific community is that the power from these mobile phone base station antennas is too low to produce health hazards as long as people are kept away from direct access to the antennas. However, current international exposure guidelines (ICNIRP) are based largely on the *thermal* effects of base station emissions. Some scientists have questioned whether there are *non thermal* effects from being exposed to low level RF such as are transmitted from mobile phone base stations. Such 'non-thermal' effects include how the actual frequencies interfere with the human brain and all other cells in the human body.

Emergency power

Fuel cell backup power systems are added to critical base stations or cell sites to provide emergency power.[15] [16]

Media

A cell tower near Thicketty, South Carolina.

Two GSM mobile phone base station towers disguised as trees in Dublin, Ireland.

A base station disguised as a palm tree in Tucson, Arizona.

Close-up of a base station antenna in Mexico City, Mexico. There are three antennas: each serves a 120-degree segment of the horizon. The microwave dish links the site with the telephone network.

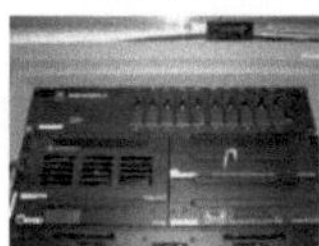

A professional rack-mount iDEN Base Radio at a Cell Site.

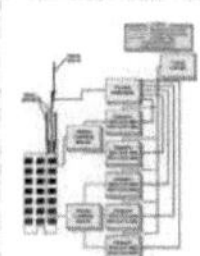

Trunked systems have groups of base stations configured as repeaters. The center blocks with frequencies in this trunked block diagram each represent a base station.

136–174 MHz US professional base station antenna examples.

WiMAX base station equipment with a sector antenna and wireless modem on top

See also

- Base transceiver station
- Mobile switching center
- Macrocell
- Microcell
- Picocell
- Femtocell
- Access point base station
- Cell site
- Cellular repeater
- Mobile phone
- Mobile phone radiation and health
- Portable phone
- Signal strength
- Audio level compression
- OpenBTS

Notes

[1] "Evaluating Regional Alternatives: Systems Design Considerations," *Planning Emergency Medical Communications: Volume 2, Local/Regional Level Planning Guide*, (Washington, D.C.: National Highway Traffic Safety Administration, US Department of Transportation, 1995) pp. 39–43.

[2] Block diagram is from: "Figure 2: Two Channel VHF Base Station," *Planning Emergency Medical Communications: Volume 2, Local/Regional Level Planning Guide*, (Washington, D.C.: National Highway Traffic Safety Administration, US Department of Transportation, 1995) pp. 42.

[3] Base stations in land mobile systems are often located at remote sites such as hilltops or water towers. Some are controlled from two or more locations. For example, a base station used to communicate with taxis may be connected to remote control consoles at both a taxi company office and an answering service for after-hours calls. The taxi company and answering service may be miles apart. Single channel base stations reduce confusion by eliminating the possibility that the wrong channel may be selected.

[4] To read more about multi-channel consoles, look at the service manual for a relatively simple console: *8-Channel Remote Console, 120, 220, 240 V AC or 12 V DC T16167 AM or BM*, 68-81021E80, (Schaumburg, Illinois: Motorola, Inc. 1980.) This is a relatively simple analog console compared to large, enterprise-level Centracom-series units.

[5] The term *RTO circuit* is legacy jargon and comes from Bell System billing terminology. RTO circuits refer to analog radio remote control and radio broadcast leased telephone circuits.

[6] For a brief discussion of remote controlled base stations, see: "Evaluating Regional Alternatives: Systems Design Considerations," *Planning Emergency Medical Communications: Volume 2, Local/Regional Level Planning Guide*, (Washington, D.C.: National Highway Traffic Safety Administration, US Department of Transportation, 1995) pp. 39–41. Tone remote controls are described in this section.

[7] Block diagram is from: "Figure 8: Bandpass Cavity/Isolator Location," *Planning Emergency Medical Communications: Volume 2, Local/Regional Level Planning Guide*, (Washington, D.C.: National Highway Traffic Safety Administration, US Department of Transportation, 1995) pp. 57.

[8] Bulleted items condensed from, "EMS Communications," "System Coordination," and "Site Engineering," in *Planning Emergency Medical Communications: Volume 2, Local/Regional Level Planning Guide*, (Washington, D.C.: National Highway Traffic Safety Administration, US Department of Transportation, 1995) pp. 10–19, 55–58.

[9] For an example of receiver specifications, see, "Table 9-3," *800 MHz Trunked Radio Request for Proposals: Public Safety Projects Office*, (Oklahoma City, Oklahoma: Oklahoma City Municipal Facilities Authority, 2000, pp. 181.

[10] A list of types of filtering used to prevent interference between equipment at the same site is included in "6.2.4 Electromagnetic Compatibility Studies," *800 MHz Trunked Radio Request for Proposals: Public Safety Projects Office*, (Oklahoma City, Oklahoma: Oklahoma City Municipal Facilities Authority, 2000, pp. 119.

[11] More detail of interference reduction equipment is provided in, "9.1.2 Base Station/Mobile Relay, 800 MHz," *800 MHz Trunked Radio Request for Proposals: Public Safety Projects Office*, (Oklahoma City, Oklahoma: Oklahoma City Municipal Facilities Authority, 2000, pp. 165–168.

[12] For example, US federal government systems are coordinated and licensed by the National Telecommunications and Information Administration. Business/Industrial Pool licenses are coordinated by the Personal Communications Industry Association (PCIA) and licensed by the Federal Communications Commission.

[13] Warren PM (2006). *IEEE C95.1-2005 "IEEE Standard for Safety Levels with Respect to Human Exposure to Radio Frequency Electromagnetic Fields, 3 kHz to 300 GHz"* Sources: US Department of Health; The World Health Organization. *IEEE.*

[14] "Electromagnetic fields and public health: Base stations and wireless technologies" (http://www.allcountries.org/health/ base_stations_and_wireless_technologies.html). *Health Topics A to Z.* . Retrieved 2011-07-01..

[15] Ballard fuel cells to power telecom backup power units for motorola (http://fr.chfca.ca/itoolkit.asp?pg=BALLARD_07132009)

[16] India telecoms to get fuel cell power (http://cleantech.com/news/3674/india-telecom-get-fuel-cells)

References

Onslow, David. "Two-Way Radio Success: How to Choose Two-Way Radios, Commercial Intercoms, and Other Wireless Communication Devices for Your Business." (http://wirelessintercomsonline.com/downloads/freebook. htm). IntercomsOnline.com. Retrieved 2008-09-07.

External links

* Occupational Safety and Health Admin. Non-Ionizing Radiation Exposure Guidelines. (http://www.osha.gov/ SLTC/radiofrequencyradiation/)

Telecommunications_network

A **telecommunications network** is a collection of terminals, links and nodes which connect to enable telecommunication between users of the terminals. Networks may use circuit switching or message switching. Each terminal in the network must have a unique address so messages or connections can be routed to the correct recipients. The collection of addresses in the network is called the address space.

The links connect the nodes together and are themselves built upon an underlying transmission network which physically pushes the message across the link.

Examples of telecommunications networks are:

* computer networks
* the Internet

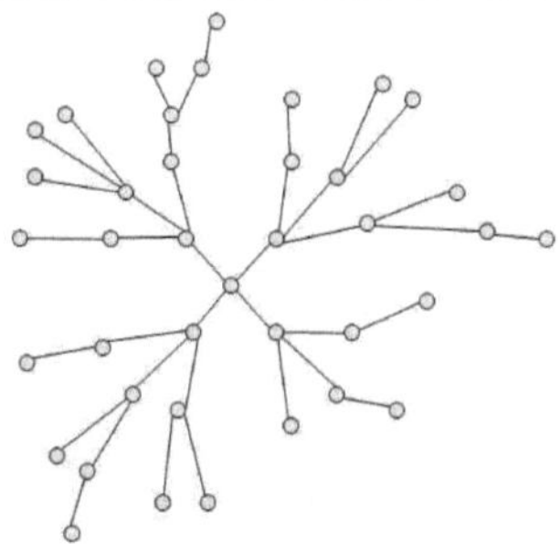

Example of how nodes may be interconnected with links to form a telecommunications network

* the telephone network
* the global Telex network
* the aeronautical ACARS network

Messages and protocols

Messages are generated by a sending terminal, then pass through the network of links and nodes until they arrive at the destination terminal. It is the job of the intermediate nodes to handle the messages and route them down the correct link toward their final destination.

These messages consist of control (or signaling) and bearer parts which can be sent together or separately. The bearer part is the actual content that the user wishes to transmit (e.g. some encoded speech, or an email) whereas the control part instructs the nodes where and possibly how the message should be routed through the network. A large number of protocols have been developed over the years to specify how each different type of telecommunication network should handle the control and bearer messages to achieve this efficiently..

Components

All telecommunication networks are made up of five basic components that are present in each network environment regardless of type or use. These basic components include terminals, telecommunications processors, telecommunications channels, computers, and telecommunications control software.

- Terminals are the starting and stopping points in any telecommunication network environment. Any input or output device that is used to transmit or receive data can be classified as a terminal component.[1]
- Telecommunications processors support data transmission and reception between terminals and computers by providing a variety of control and support functions. (i.e. convert data from digital to analog and back) [1]
- Telecommunications channels are the way by which data is transmitted and received. Telecommunication channels are created through a variety of media of which the most popular include copper wires and coaxial cables (structured cabling). Fiber-optic cables are increasingly used to bring faster and more robust connections to businesses and homes.[1]
- In a telecommunication environment computers are connected through media to perform their communication assignments.[1]
- Telecommunications control software is present on all networked computers and is responsible for controlling network activities and functionality.[1]

Early networks were built without computers, but late in the 20th century their switching centers were computerized or the networks replaced with computer networks.

Network structure

In general, every telecommunications network conceptually consists of three parts, or planes (so called because they can be thought of as being, and often are, separate overlay networks):

- The control plane carries control information (also known as signalling).
- The data plane or user plane or bearer plane carries the network's users traffic.
- The management plane carries the operations and administration traffic required for network management.

Example: the TCP/IP data network

The data network is used extensively throughout the world to connect individuals and organizations. Data networks can be connected to allow users seamless access to resources that are hosted outside of the particular provider they are connected to. The Internet is the best example of many data networks from different organizations all operating under a single address space.

Terminals attached to TCP/IP networks are addressed using IP addresses. There are different types of IP address, but the most common is IP Version 4. Each unique address consists of 4 integers between 0 and 255, usually separated by dots when written down, e.g. 82.131.34.56.

TCP/IP are the fundamental protocols that provide the control and routing of messages across the data network. There are many different network structures that TCP/IP can be used across to efficiently route messages, for example:

- wide area networks (WAN)
- metropolitan area networks (MAN)
- local area networks (LAN)
- campus area networks (CAN)
- virtual private networks (VPN)

There are three features that differentiate MANs from LANs or WANs:

1. The area of the network size is between LANs and WANs. The MAN will have a physical area between 5 and 50 km in diameter.[2]

2. MANs do not generally belong to a single organization. The equipment that interconnects the network, the links, and the MAN itself are often owned by an association or a network provider that provides or leases the service to others.[2]

3. A MAN is a means for sharing resources at high speeds within the network. It often provides connections to WAN networks for access to resources outside the scope of the MAN.[2]

See also

- Active networking
- Access network
- Core network
- Network analyzer
- Coverage (telecommunication)
- Double-ended synchronization
- Federation (information technology)
- MVNE
- MVNO
- Network node
- Nanoscale network
- Network model
- Optical fiber
- Submarine communications cable
- Optical mesh network
- Telecommunications Industry Association

References

[1] O'Brien, J. A. & Marakas, G. M. (2008). Management Information Systems. New York: McGraw-Hill Irwin.

[2] http://www.erg.abdn.ac.uk/users/gorry/course/intro-pages/man.html

Footprint_(satellite)

The **footprint** of a communications satellite is the ground area that its transponders offer coverage, and determines the satellite dish diameter required to receive each transponder's signal. There is usually a different map for each transponder (or group of transponders) as each may be aimed to cover different areas of the ground.

Footprint maps usually show either the estimated minimum satellite dish diameter required or the signal strength in each area measured in dBW.

External links

- Links to fleet information and footprints [1] from SES.
- Links to interactive maps [2] from Intelsat for their fleet of satellites.
- Links to interactive maps [3] for SES World Skies's fleet of satellites.
- Link to maps [4] for Russian Satellite Communications Company satellites
- Link to satellite footprints [5] from SatBeams for Geostationary satellites

An example of an elliptical footprint with a reception area of Germany, Austria and Switzerland. The ellipses indicate the necessary antenna diameter for receiving in cm.

References

[1] http://www.ses.com/4232618/fleet-coverage

[2] http://www.intelsat.com/network/satellite/index.asp

[3] http://www.sesworldskiesfleetmap.com/pages/home.aspx

[4] http://eng.rscc.ru/171/197/

[5] http://www.satbeams.com/footprints

Article Sources and Contributors

Coverage_(telecommunication) *Source*: http://en.wikipedia.org/w/index.php?title=Coverage_%28telecommunication%29 *Contributors*: Altermike, AugustoRomero, Chillysnow, DirkvdM, Doris Camire, Fordfusion2010, Hu12, Jim.henderson, Mac, MarkPos, Nopetro, Nudecline, Nukeless, Radagast83, TastyPoutine, Tomi T Ahonen, Vegaswikian, Woohookitty, 3 anonymous edits

Broadcasting *Source*: http://en.wikipedia.org/w/index.php?title=Broadcasting *Contributors*: 16@r, A. B., A.Abdel-Rahim, ABF, Adashiel, Alai, Alansohn, Alasdair, Alexv7255, Allyn, Altenmann, Andyjsmith, Angela, Angmering, Anna Lincoln, Anwar saadat, Art LaPella, Ask123, Aunt Entropy, BarkingFish, Bartolomas214, Bdragon, Belovedfreak, Ben Ben, BenFrantzDale, Bernelis, Bhuston, Bing98748109347, Binksternet, Birtitia, Blackwellmas229, Blehfu, Bluedisk, Bob cat, Bobblehead, Bobblewik, Bobby131313, Boffob, BowmanJason, Brian0918, Butler david, Cab88, Cacophony, Caggegi229, CambridgeBayWeather, Capricorn42, Catdude, CatherineMunro, Cavenba, Chamal N, CharlieZeb, Chasingsol, Chotisornmas214, Christian List, Chuenprayothmas214, Cindy141, Ckape, Clegs, Cocoloco, Coinchon, Colleenthegreat, ComicMasta, CommonsDelinker, Coster34, Cowpip, Creator-bz, Crispy park, D2s, DDerby, Dailyvoid, Dale Arnett, Darren Jowalsen, David Kernow, Denelson83, Dennis Kussinich 08, Deville, Dnyhagen, DocWatson42, Dp67, Drakes are abound, Dropbuilt1234, Dzulkk, Edison, Edwina Storie, Enochlau, Enterpoint, EoGuy, Erpel13, Erzahler, Euchiasmus, Everyking, Fahrenheit451, Falcon8765, FrummerThanThou, Funandtrvl, Fyyer, GABaker, GEBStgo, Gadfium, GarnetRChaney, Gavin Starks, Gioto, Giraffedata, Gladys j cortez, Glenn, Goplat, Grammar conquistador, Gscshoyru, Guaka, Hairy Dude, Hallenrm, HardyMAS229, Harris-Grad, Harryzilber, Harumphy, Hbent, Healy229, Hertz1888, Iandstanley, Idleguy, Ilyarmas214, IslandGyrl, Ismailmas214, JHeinonen, Jacobmas229, Jakro64, JamesBWatson, Janneok, JeLuF, Jerryseinfeld, Jerzy, JoaoRicardo, Johnuniq, JordoCo, Joseph Solis in Australia, Julianp, Junglecat, KFP, KGun10, Kallemax, Kashi0341, Kbrose, Keno, Khoikhoi, Kilo-Lima, Klokie, Krwells, Lauramartz, Laycockmas229, Lee M, Leimas229, Lesterxyz, Levineps, LorenzoB, Lotje, Luna Santin, MDGx, MGlosenger, MMuzammils, Mandylyn, Mange01, Martin-C, Mas214Kapinga, Mattbr, Mattimeeleo, Mayawi, McDonaldmas214, Mclaughlinmas214, MichaelJanich, Mike Russell, Mikeblas, Milenkovic214, Minesweeper, Miniapolis, Mintguy, MoarNoir-Mas229, Momet, Monaco Kati, MonstaPro, MrOllie, Mssetiadi, Mulad, Mushroom, Mwanner, Mxn, N24p, Naddy, NawlinWiki, Nedim Ardoğa, Neutralhomer, Nintendude, Noisy, Nommonomanac, Northamerica1000, Ohnoitsjamie, Oicumayberight, OlEnglish, Oli Filth, Olisssr, OnBeyondZebrax, Patrick, Pattersonmas229, Paul Barlow, Paul Benjamin Austin, Paul Erik, Pearle, Pelago, Peoplesyak, Picapica, Pichu0102, Pigmingo, Pit-yacker, Prattflora, Pro bug catcher, Qxz, R'n'B, Radiojon, Raja229, Raven4x4x, Reconsider the static, RedWolf, Redvers, Requestion, RexNL, Rich Farmbrough, Rjensen, Rob.au, Robin Patterson, Rocketgoat, Rror, Rtdixon86, Rumiton, Ryan1 walker, S.M.Latif Shahid, SIGURD42, Saforrest, Sagaciousuk, Schneelocke, SchreiberBike, SebastianBreier, Sfeagles5, Shadowjams, Shalom Yechiel, Shoreranger, Sigma 7, Sinblox, Skizzik, Smokizzy, SpLoT, Spangineer, Spliced, SqueakBox, Ssr, SudoGhost, Swalker2000, Synergy radio, TFOWR, TMillerCA, TUF-KAT, Taylormas229, Tcooling, TeaDrinker, TheObtuseAngleOfDoom, Thomas Blomberg, Thruston, Thue, Tide rolls, Tiwifey1, Tomas e, Tomwalden, Toussaint, Triktrak, Tsomas214, Twas Now, UncivilFire, Unreal7, Vicarious, Vidshow, Violetriga, Walkerma, Wayiran, Wayland, Wcquidditch, Wik, Wiki Roxor, Wikiklrsc, Willsmith, Winger84, Wmahan, Woohookitty, Wp, Wysprgr2005, Xtzou, Yanksox, Yansa, Yarnalgo, Zealotgi, Zoicon5, Zondor, ملاع بوبحم, 299 anonymous edits

Radio_broadcasting *Source*: http://en.wikipedia.org/w/index.php?title=Radio_broadcasting *Contributors*: 121a0012, 16@r, A. B., A.M.962, Aaronrg, Abe Lincoln, Allreet, Am088, Andres, Andrewpmk, AndreyMavlyanov, Anna Frodesiak, Annelid, Anwar saadat, Armistej, Barrett101, Bearcat, Beatgr, BecauseWhy?, Ben Ward, Benjamin Mako Hill, Bigmorr, Bitplane, Blainster, Boffob, Bonadea, Brakea, Bring back ILR, Bull-Doser, CambridgeBayWeather, Canadian Scouter, Cansecon, Chessphoon, ChrisGriswold, Civil Engineer III, Cmh, Cohesion, CommonsDelinker, Correogsk, Coster34, Cowpip, DH1992, Daniel C, DavidMendoza, Deisenbe, Denelson83, DenisRS, Dicksinthebag, Dirkbike, Dravecky, Dudesleeper, ESkog, Edderso, Edward, El C, ElSaxo, Epolk, Escloupere, Excirial, Fanatix, Fishyghost, Fmi7323104, Fratrep, Fred32323232323232, Funandtrvl, Gameplaya888, Gene Nygaard, Georgia guy, Gf uip, GrayFullbuster, Gtstricky, Gurch, Halmstad, Harryboyles, Harryzilber, Helpper, Hertz1888, Hotcrocodile, Hotpotato1234, Hulmem, IPSOS, Iamknowledged, Ilyaroz, Infrogmation, Ishan 2008, Isnow, JA.Davidson, JCRansom, JForget, JPG-GR, Jacooks, Jaknouse, Jakro64, Jdavidb, Jlin, Joedeshon, Jonnny, JordoCo, Jpgordon, Judicatus, JustAGal, Katecummings, Kbrose, Kchishol1970, Kcordina, Kharker, Kjpoconnor, Kozuch, Krash, Kyng, LFaraone, LaMenta3, Larrymcp, Liverlife, Lommer, Lotje, Majorxp, MariusG, Mav, Mazin07, Mbessey, Mhking, Mintguy, Modest Genius, Mohamedishakhowth, Moline670, Moreau1, MrRadioGuy, Mulad, Mwanner, MyFavoriteMartin, Neelix, Newsjessore, Nick C, Nightkey, Nposs, Ojigiri, Omkar1234, Paigntonuk, Penx, Persiana, Piano non troppo, Pillus, Pjoef, Puattmas, Qwertylurker, RadioFan, RadioTheodric, Radiobill, Radiojon, RafaAzevedo, Random11mason, Randwicked, Reddi, Redvers, Rich257, Rjensen, Robert Skyhawk, Robert1947, Ronvelig, Rrburke, Rwagoner, Rydel, SEWilco, SNIyer12, SSBDelphiki, Scottfisher, Seaphoto, Seowonk, Sergioledesma, Shalom Yechiel, Sherbrooke, Simply south, Sjb90, Slady, Slightsmile, Sprocket, Stemonitis, Stormie, Svick, SwisterTwister, Swtpc6800, Tannin, Tedder, Tedernst, The Evil IP address, The Thing That Should Not Be, Thejpmshow, TiCPU, Tohd8BohaithuGh1, Toussaint, Transent, URBMark, Unara, Uncle Dick, Vanwhistler, Vcelloho, Veinor, Wfeidt, Woohookitty, Wtshymanski, Xombie, Y control, Z388, Zzuuzz, 265 anonymous edits

Telecommunication *Source*: http://en.wikipedia.org/w/index.php?title=Telecommunication *Contributors*: 11cookeaw1, 130.94.122.xxx, 16@r, 213.121.101.xxx, 2over0, 4pq1injbok, 4twenty42o, 5 albert square, A New Nation, A. B., A5b, ARUNKUMAR P.R, Achowat, Addshore, Aden Scott, Admiral Roo, Ahoerstemeier, Ajshm, Alansohn, Aldie, Alrino, Anareon, Andrew Hartford, Antandrus, Anth12, Ap, Appraiser, ArielGold, Art LaPella, Ashishkulshreshtha, Asocall, Aulman, BD2412, Basangbur, Batmanand, Beetstra, Bert490, Betsumei, Bhavin105, Bkmays, Blue520, Bluelion, Bluemask, Blycroft, Bobblewik, Bobo192, Bogatabeav, Borislav, Borofkin, Bowmanmas229, Bradeos Graphon, Bridget Huntley, Brittanyab, Brolsma, Bryan Derksen, Bsimmons666, Bucketsofg, Bushsf, Businessphonesystems, Bwhack, CALR, CRGreathouse, CambridgeBayWeather, Canaima, Card, Casey Abell, Cedars, Ceyockey, Ched Davis, Chimpex, Chris857, Chuunen Baka, ClovisPt, Colbry53, Colonies Chris, CommonsDelinker, Conversion script, Coreyxks, CosineKitty, Courcelles, Cquan, Crusoe8181, Curps, CyclePat, DMcMPO11AAUK, DR04, DVdm, David Jordan, DavidWBrooks, Dawnseeker2000, Dcljr, Deglr6328, Demicx, DerHexer, Dicklyon, Discospinster, Download, Draksis314, DynamoDegsy, Edcolins, Edward, El C, Elagoutova, Elassint, Elm-39, Erkan Yilmaz, Ethan01, Europaandio, Ezhuttukari, FCYTravis, FF2010, Favonian, Ffxaddict899, Finalius, From-cary, Funandtrvl, Fvw, Fæ, GEBStgo, Gaga.vaa, Gaianauta, Gail, Galoubet, GcSwRhIc, Georgiamonet, Gggh, Giftlite, Glenn, Gloop, Gogo Dodo, GoingBatty, GorillaWarfare, GraemeL, Graham87, Grammargal, Guaka, Gururaju, Haakon, Hadal, Hadiyana, Harryboyles, Harryzilber, Haylstorms, Hcberkowitz, Hdorren, HelinaZ, Herodotos, Heron, Hittman627, Hmains, Hughdbrown, Hugsforsale, Hulagutten, Hydnjo, IGeMiNix, IMarc89, IdleUser, Indosauros, Inkling, Insanity Incarnate, Intgr, InverseHypercube, Ioverka, Iridescent, It Is Me Here, J.delanoy, JD554, JDP90, Jaxl, Jayeshtula, Jeremyjf22, Jerryseinfeld, Jeysaba, Jim.henderson, Jim1138, JoeSmack, John of Reading, John254, JohnCD, JohnGray, JohnTechnologist, Johnpseudo, Johnuniq, Jose77, Joseph Solis in Australia, Julie Deanna, Juzaily, K12u, Kaeh4, Katherine, Kayau, Kbdank71, Keilana, KelleyCook, Kelly Martin, Kerotan, Kjkarthikmaddy, Kjoonlee, Kjoseph7777, Kkm010, Kku, Koavf, Kozuch, Krakfly, Kristine.clara, Kutulus, LIVPAT, Lakhim, Lars Washington, Leszek Jańczuk, Light current, Lightmouse, LoKiLeCh, Logan, LorenzoB, Lotje, Lperez2029, Luna Santin, Lxdbsn, M1ss1ontomars2k4, MCI telcomm, MER-C, MacMed, Madhero88, Mahjongg, Majorly, Makru, Malleus Fatuorum, Marc Spoddle, MarkSweep, Martarius, Matilda, Matthewedwards, Maurreen, Mav, Maximus Rex, Mblaze, Mboverload, McSly, Mendaliv, Mendel, Mentifisto, Merlinsorca, Mezzaluna, Mhughes38, Michael Hardy, Mikeblas, Milesdowsett, Mindmatrix, Mion, Misto, Mitsuhirato, Mlewis000, Mnativesacl, Modulatum, Mohan0704, Moonriddengirl, Mr Myson, Mr Stephen, MrOllie, MrSomeone, Mrgates, Munford, Murali intl, Mīthrandir, N2e, Nasa-verve, Nazi 2007, Netalarm, Neutrality, Nikai, NocturnalA6 2.7, Northamerica1000, Obornp, Ohad.cohen, Ohnoitsjamie, Oicumayberight, OlEnglish, Oli Filth, Olivier, Once in a Blue Moon, OrgasGirl, Ossguy, PFHLai, PJY, Paul J Wayman, Pavel Vozenilek, Payxystaxna, Persian Poet Gal, Pharaoh of the Wizards, PhilHibbs, PhilipO, Phoenix2, Piano non troppo, Piotrus, Ppntori, Pseudomonas, Pugliavi, Pvineet131, Pwarrior, Qedu, Qsa5kn, Quarl, R'n'B, RW Marloe, Radavidson, Radishes, Ralesk, Rangoon11, Rapaporta, Rasmus Faber, Ratiocinate, Raven in Orbit, Razorflame, Readiwip, RedWolf, Reddi, Remi0o, RexNL, Rfc1394, Riana, Rich Farmbrough, Richardpitt, Rick Block, Rick Sidwell, Rifasj123, Rillian, Rjstott, Rjwilmsi, Robbyyy, Romann, Rory096, Rrburke, Rsayles, RuM, Rwwww, SD5, SEWilco, Saligron, Sam Blacketer, Samsam.yh, SandyGeorgia, Sannse, SansSanity, Sbyrnes321, Scarian, Schproject, Sdsds, Sepersann, Setherson, Shadowjams, ShakingSpirit, Shangrilaista, Shenme, Silly rabbit, Skater, Skier Dude, SkylineEvo, Smartishkindaguy, Smasafy, Smyth, Snafflekid, Solidice190, Solveforce, Sonjaaa, Spatch, SpecMode, Special-T, Spitfire19, Splash, Sprinklezz, Srleffler, Srobak, Steeev, Stenson jack, Stephenw77, Stewartadcock, Stirling Newberry, Studerby, SudoGhost, SunCreator, Svick, Swamynathan007, Sweetpoet, Symane, Sysiphe, THEN WHO WAS PHONE?, Tabletop, TastyPoutine, Telecom.portal, Telecomman, Tellyaddict, Template namespace initialisation script, TenOfAllTrades, Terrx, The Anome, The Cunctator, The Thing That Should Not Be, The Transhumanist (AWB), TheGrimReaper NS, Thrissel, Timsheridan, Tony1, Totel, Tpbradbury, Tregoweth, Trevor MacInnis, Tristanb, Trusilver, Tsiuser09, Tuxa, Tyrenius, Tyw7, Unyoyega, Usingha, V4nd3r, Valeria70, Vasu99a, Vegaswikian, Veinor, Venya, Verkhovensky, Vespristiano, Vihar7, Violetriga, Wa3frp, Wafulz, WaltBusterkeys, WannabeAmatureHistorian, Wavelength, Whiskers9, Wik, Wiki2contrib, WikiLaurent, Wikiklrsc, Wikipelli, Will, Wizardist, Woohookitty, Wtmitchell, Wutsje, XxTimberlakexx, Zedh, ZeroOne, Милан Јелисавчић, ТимофейЛееСуда, אמסי'סה123, ה־תעבם, םרד־, 886 anonymous edits

Coverage_map *Source*: http://en.wikipedia.org/w/index.php?title=Coverage_map *Contributors*: Benjamin Mako Hill, Chillysnow, Estrategy, Finefox771, Jim.henderson, JustinSmith, MGlosenger, MarkPos, Numen, RTG, Szyslak, TejasDiscipulus2, TheGerm, Vegaswikian, WVhybrid, 2 anonymous edits

Orography *Source*: http://en.wikipedia.org/w/index.php?title=Orography *Contributors*: Alansohn, Altenmann, Brambleshire, Cdouglas, Chzz, Cuvette, Darwinek, Dougher, Frokor, Frosted14, Gaius Cornelius, Ian Spackman, Ingolfson, Jeroen, KPH2293, Liamdaly620, Look2See1, Mackinaw, MrBell, MrMambo, Mzajac, Nudecline, PhilMacD, Random account 47, Redvers, SchuminWeb, Sdaines, Semolo75, Sp33dyphil, Spiritia, Stan Shebs, Txomin, Whpq, 42 anonymous edits

Multi-storey_car_park *Source*: http://en.wikipedia.org/w/index.php?title=Multi-storey_car_park *Contributors*: Aadh, Academic Challenger, Acroterion, Ahunt, Alansohn, Algont, Alphachimp, Andy Janata, ArnoldReinhold, BMF81, Bdiscoe, Beetstra, CBM, CalderOliver, Carguy.Barnes, Cleared as filed, Cohesion, Coolhawks88, Dah31, Daniel Christensen, David Shay, DeadEyeArrow, Delirium, Demize, Dough4872, Dougweller, Dream out loud, ESkog, Edal, Ekashp, Eliyak, Everything counts, Fram, Golfandme, GorgeCustersSabre, Grutness, Haham hanuka, Halfalah, Hi878, Hokanomono, Hu, I Dig Cars, Interiot, Islander, Jackehammond, Jackel, Jesster79, Jim.henderson, Jnelson09, Johnmartins, Jrleighton, Kalmia, Kinkladze28, Kjlewis, LaptopMad777, Lee M, LeeGilliss, Lightmouse, Mailer diablo, Maxxies, Mets501, Michael Hardy, Millbrooky, Milnivlek, Mitsukai, Mwaldo, Neutrality, Nhankey, Ohconfucius, Ordew, PPdd, Parkingfacebook, Paul W, Pcpcpc, Phurley93, Picapica, Plasticup, Purgatory Fubar, PvsKllKsVp, Radiojon, Remag Kee, Risker, Rklawton, Rory096, Roselli78fp, RottweilerCS, Saros136, Shadowlink1014, Snillet, Sociotard, Sumsum2010, TVBZ28, TerriersFan, The Anome, TheTruthiness, TheoThoughts, Uagehry456, UberMan5000, Urbaneddie, Vegaswikian, Wasted Time R, Wiki Wikardo, Woohookitty, Wtshymanski, Zegoma beach, 134 anonymous edits

Base_station *Source*: http://en.wikipedia.org/w/index.php?title=Base_station *Contributors*: 16@r, Anikingos, Animal00, BD2412, Bellenion, Bigglescat, BlackIceNRW, Blorg, Bobo192, Bvob, Chekovf, Chris the speller, ChrisUK, Cncs wikipedia, David Jordan, Dennis TYCC, Echosmoke, EleganceRevived, Enigmaman, Fallschirmjäger, Glenn, Harryzilber, Helt, Irishguy, JKofoed,

JeffreyBradley, Jim.henderson, JohnTechnologist, KPF, Karl Naylor, Kotra, Kozuch, Kwnd, KyleAndMelissa22, Marek69, Minnaert, Mion, Modster, Mozzerati, My76Strat, Navy Blue, Neckelmann, Nobodyinpart, Petri Krohn, Pptudela, R'n'B, Radu - Eosif Mihailescu, Rsabbatini, Sam Hocevar, Silvonen, Skazella, Slashme, Squiggleslash, Stannered, StephanieM, Sv1xv, TastyPoutine, Thomas Ludwig, Thumperward, Tomas e, Txuspe, VCA, Vegaswikian, Wtshymanski, Вильдан, הרישבבוח, 56 anonymous edits

Telecommunications_network *Source*: http://en.wikipedia.org/w/index.php?title=Telecommunications_network *Contributors*: 802geek, Aarky, Aldie, Amairis, Amysass5, Amyso5, Apparition11, B4hand, BRUTE, Bensaccount, Brittanyab, Bushsf, Chowbok, ChrisUK, Christian75, Denoir, Dgtsyb, EoGuy, Evans1982, Firsfron, Foxy55, Gilgar, Grigio60, Harryzilber, Hu12, Itusg15q4user, J.delanoy, JanCeuleers, Jflabourdette, Jim.henderson, JonHarder, JordoCo, Jpbowen, Kingpin13, Loftenter, MMuzammils, Makru, Mange01, Mlewis000, Nokkosukko, Nubiatech, NuclearWinner, Nudecline, Nurg, Petri Krohn, RHaworth, SamJohnston, Scientus, Signalhead, Someguy1221, Srinivasasha, Srleffler, Steven.dai, Storm Rider, TheAMmollusc, Thedon1, Thryduulf, Tmn, Turzh, Wasbeer, WriteManWriting, Y2karmagedon, 72 anonymous edits

Footprint_(satellite) *Source*: http://en.wikipedia.org/w/index.php?title=Footprint_%28satellite%29 *Contributors*: Abdull, Amalas, Andy29, CanadianLinuxUser, Chillysnow, ElectraFlarefire, Finlay McWalter, Gail, Jamoche, Jmccormac, Kerotan, M jurrens, Mikebar, Naddy, Niteowlneils, Nukeless, RJFJR, Sam8, Satbuff, Satfootprint, SchreiberBike, TobyDZ, Toffile, Uwe W., WDGraham, West London Dweller, Whosasking, 24 anonymous edits

Image Sources, Licenses and Contributors

GNU Free Documentation License Version 1.2, November 2002 Copyright (C) 2000,2001,2002 Free Software Foundation, Inc. 59 Temple Place, Suite 330, Boston, MA 02111-1307 USA Everyone is permitted to copy and distribute verbatim copies of this license document, but changing it is not allowed.

0. PREAMBLE

The purpose of this License is to make a manual, textbook, or other functional and useful document "free" in the sense of freedom: to assure everyone the effective freedom to copy and redistribute it, with or without modifying it, either commercially or noncommercially. Secondarily, this License preserves for the author and publisher a way to get credit for their work, while not being considered responsible for modifications made by others. This License is a kind of "copyleft", which means that derivative works of the document must themselves be free in the same sense. It complements the GNU General Public License, which is a copyleft license designed for free software. We have designed this License in order to use it for manuals for free software, because free software needs free documentation: a free program should come with manuals providing the same freedoms that the software does. But this License is not limited to software manuals; it can be used for any textual work, regardless of subject matter or whether it is published as a printed book. We recommend this License principally for works whose purpose is instruction or reference.

1. APPLICABILITY AND DEFINITIONS

This License applies to any manual or other work, in any medium, that contains a notice placed by the copyright holder saying it can be distributed under the terms of this License. Such a notice grants a world-wide, royalty-free license, unlimited in duration, to use that work under the conditions stated herein. The "Document", below, refers to any such manual or work. Any member of the public is a licensee, and is addressed as "you". You accept the license if you copy, modify or distribute the work in a way requiring permission under copyright law. A "Modified Version" of the Document means any work containing the Document or a portion of it, either copied verbatim, or with modifications and/or translated into another language. A "Secondary Section" is a named appendix or a front-matter section of the Document that deals exclusively with the relationship of the publishers or authors of the Document to the Document's overall subject (or to related matters) and contains nothing that could fall directly within that overall subject. (Thus, if the Document is in part a textbook of mathematics, a Secondary Section may not explain any mathematics.) The relationship could be a matter of historical connection with the subject or with related matters, or of legal, commercial, philosophical, ethical or political position regarding them. The "Invariant Sections" are certain Secondary Sections whose titles are designated, as those of Invariant Sections, in the notice that says that the Document is released under this License. If a section does not fit the above definition of Secondary then it is not allowed to be designated as Invariant. The Document may contain zero Invariant Sections. If the Document does not identify any Invariant Sections then there are none. The "Cover Texts" are certain short passages of text that are listed, as Front-Cover Texts or Back-Cover Texts, in the notice that says that the Document is released under this License. A Front-Cover Text may be at most 5 words, and a Back-Cover Text may be at most 25 words. A "Transparent" copy of the Document means a machine-readable copy, represented in a format whose specification is available to the general public, that is suitable for revising the document straightforwardly with generic text editors or (for images composed of pixels) generic paint programs or (for drawings) some widely available drawing editor, and that is suitable for input to text formatters or for automatic translation to a variety of formats suitable for input to text formatters. A copy made in an otherwise Transparent file format whose markup, or absence of markup, has been arranged to thwart or discourage subsequent modification by readers is not Transparent. An image format is not Transparent if used for any substantial amount of text. A copy that is not "Transparent" is called "Opaque". Examples of suitable formats for Transparent copies include plain ASCII without markup, Texinfo input format, LaTeX input format, SGML or XML using a publicly available DTD, and standard-conforming simple HTML, PostScript or PDF designed for human modification. Examples of transparent image formats include PNG, XCF and JPG. Opaque formats include proprietary formats that can be read and edited only by proprietary word processors, SGML or XML for which the DTD and/or processing tools are not generally available, and the machine-generated HTML, PostScript or PDF produced by some word processors for output purposes only. The "Title Page" means, for a printed book, the title page itself, plus such following pages as are needed to hold, legibly, the material this License requires to appear in the title page. For works in formats which do not have any title page as such, "Title Page" means the text near the most prominent appearance of the work's title, preceding the beginning of the body of the text. A section "Entitled XYZ" means a named subunit of the Document whose title either is precisely XYZ or contains XYZ in parentheses following text that translates XYZ in another language. (Here XYZ stands for a specific section name mentioned below, such as "Acknowledgements", "Dedications", "Endorsements", or "History".) To "Preserve the Title" of such a section when you modify the Document means that it remains a section "Entitled XYZ" according to this definition. The Document may include Warranty Disclaimers next to the notice which states that this License applies to the Document. These Warranty Disclaimers are considered to be included by reference in this License, but only as regards disclaiming warranties: any other implication that these Warranty Disclaimers may have is void and has no effect on the meaning of this License.

2. VERBATIM COPYING

You may copy and distribute the Document in any medium, either commercially or noncommercially, provided that this License, the copyright notices, and the license notice saying this License applies to the Document are reproduced in all copies, and that you add no other conditions whatsoever to those of this License. You may not use technical measures to obstruct or control the reading or further copying of the copies you make or distribute. However, you may accept compensation in exchange for copies. If you distribute a large enough number of copies you must also follow the conditions in section 3. You may also lend copies, under the same conditions stated above, and you may publicly display copies.

3. COPYING IN QUANTITY

If you publish printed copies (or copies in media that commonly have printed covers) of the Document, numbering more than 100, and the Document's license notice requires Cover Texts, you must enclose the copies in covers that carry, clearly and legibly, all these Cover Texts: Front-Cover Texts on the front cover, and Back-Cover Texts on the back cover. Both covers must also clearly and legibly identify you as the publisher of these copies. The front cover must present the full title with all words of the title equally prominent and visible. You may add other material on the covers in addition. Copying with changes limited to the covers, as long as they preserve the title of the Document and satisfy these conditions, can be treated as verbatim copying in other respects. If the required texts for either cover are too voluminous to fit legibly, you should put the first ones listed (as many as fit reasonably) on the actual cover, and continue the rest onto adjacent pages. If you publish or distribute Opaque copies of the Document numbering more than 100, you must either include a machine-readable Transparent copy along with each Opaque copy, or state in or with each Opaque copy a computer-network location from which the general network-using public has access to download using public-standard network protocols a complete Transparent copy of the Document, free of added material. If you use the latter option, you must take reasonably prudent steps, when you begin distribution of Opaque copies in quantity, to ensure that this Transparent copy will remain thus accessible at the stated location until at least one year after the last time you distribute an Opaque copy (directly or through your agents or retailers) of that edition to the public. It is requested, but not required, that you contact the authors of the Document well before redistributing any large number of copies, to give them a chance to provide you with an updated version of the Document.

4. MODIFICATIONS

You may copy and distribute a Modified Version of the Document under the conditions of sections 2 and 3 above, provided that you release the Modified Version under precisely this License, with the Modified Version filling the role of the Document, thus licensing distribution and modification of the Modified Version to whoever possesses a copy of it. In addition, you must do these things in the Modified Version: A. Use in the Title Page (and on the covers, if any) a title distinct from that of the Document, and from those of previous versions (which should, if there were any, be listed in the History section of the Document). You may use the same title as a previous version if the original publisher of that version gives permission. B. List on the Title Page, as authors, one or more persons or entities responsible for authorship of the modifications in the Modified Version, together with at least five of the principal authors of the Document (all of its principal authors, if it has fewer than five), unless they release you from this requirement. C. State on the Title page the name of the publisher of the Modified Version, as the publisher. D. Preserve all the copyright notices of the Document. E. Add an appropriate copyright notice for your modifications adjacent to the other copyright notices. F. Include, immediately after the copyright notices, a license notice giving the public permission to use the Modified Version under the terms of this License, in the form shown in the Addendum below. G. Preserve in that license notice the full lists of Invariant Sections and required Cover Texts given in the Document's license notice. H. Include an unaltered copy of this License. I. Preserve the section Entitled "History", Preserve its Title, and add to it an item stating at least the title, year, new authors, and publisher of the Modified Version as given on the Title Page. If there is no section Entitled "History" in the Document, create one stating the title, year, authors, and publisher of the Document as given on its Title Page, then add an item describing the Modified Version as stated in the previous sentence. J. Preserve the network location, if any, given in the Document for public access to a Transparent copy of the Document, and likewise the network locations given in the Document for previous versions it was based on. These may be placed in the "History" section. You may omit a network location for a work that was published at least four years before the Document itself, or if the original publisher of the version it refers to gives permission. K. For any section Entitled "Acknowledgements" or "Dedications", Preserve the Title of the section, and preserve in the section all the substance and tone of each of the contributor acknowledgements and/or dedications given therein. L. Preserve all the Invariant Sections of the Document, unaltered in their text and in their titles. Section numbers or the equivalent are not considered part of the section titles. M. Delete any section Entitled "Endorsements". Such a section may not be included in the Modified Version. N. Do not retitle any existing section to be Entitled "Endorsements" or to conflict in title with any Invariant Section. O. Preserve any Warranty Disclaimers. If the Modified Version includes new front-matter sections or appendices that qualify as Secondary Sections and contain no material copied from the Document, you may at your option designate some or all of these sections as invariant. To do this, add their titles to the list of Invariant Sections in the Modified Version's license notice. These titles must be distinct from any other section titles. You may add a section Entitled "Endorsements", provided it contains nothing but endorsements of your Modified Version by various parties--for example, statements of peer review or that the text has been approved by an organization as the authoritative definition of a standard. You may add a passage of up to five words as a Front-Cover Text, and a passage of up to 25 words as a Back-Cover Text, to the end of the list of Cover Texts in the Modified Version. Only one passage of Front-Cover Text and one of Back-Cover Text may be added by (or through arrangements made by) any one entity. If the Document already includes a cover text for the same cover, previously added by you or by arrangement made by the same entity you are acting on behalf of, you may not add another; but you may replace the old one, on explicit permission from the previous publisher that added the old one. The author(s) and publisher(s) of the Document do not by this License give permission to use their names for publicity for or to assert or imply endorsement of any Modified Version.

5. COMBINING DOCUMENTS

You may combine the Document with other documents released under this License, under the terms defined in section 4 above for modified versions, provided that you include in the combination all of the Invariant Sections of all of the original documents, unmodified, and list them all as Invariant Sections of your combined work in its license notice, and that you preserve all their Warranty Disclaimers. The combined work need only contain one copy of this License, and multiple identical Invariant Sections may be replaced with a single copy. If there are multiple Invariant Sections with the same name but different contents, make the title of each such section unique by adding at the end of it, in parentheses, the name of the original author or publisher of that section if known, or else a unique number. Make the same adjustment to the section titles in the list of Invariant Sections in the license notice of the combined work. In the combination, you must combine any sections Entitled "History" in the various original documents, forming one section Entitled "History"; likewise combine any sections Entitled "Acknowledgements", and any sections Entitled "Dedications". You must delete all sections Entitled "Endorsements".

6. COLLECTIONS OF DOCUMENTS

You may make a collection consisting of the Document and other documents released under this License, and replace the individual copies of this License in the various documents with a single copy that is included in the collection, provided that you follow the rules of this License for verbatim copying of each of the documents in all other respects. You may extract a single document from such a collection, and distribute it individually under this License, provided you insert a copy of this License into the extracted document, and follow this License in all other respects regarding verbatim copying of that document.

7. AGGREGATION WITH INDEPENDENT WORKS

A compilation of the Document or its derivatives with other separate and independent documents or works, in or on a volume of a storage or distribution medium, is called an "aggregate" if the copyright resulting from the compilation is not used to limit the legal rights of the compilation's users beyond what the individual works permit. When the Document is included in an aggregate, this License does not apply to the other works in the aggregate which are not themselves derivative works of the Document. If the Cover Text requirement of section 3 is applicable to these copies of the Document, then if the Document is less than one half of the entire aggregate, the Document's Cover Texts may be placed on covers that bracket the Document within the aggregate, or the electronic equivalent of covers if the Document is in electronic form. Otherwise they must appear on printed covers that bracket the whole aggregate.

8. TRANSLATION

Translation is considered a kind of modification, so you may distribute translations of the Document under the terms of section 4. Replacing Invariant Sections with translations requires special permission from their copyright holders, but you may include translations of some or all Invariant Sections in addition to the original versions of these Invariant Sections. You may include a translation of this License, and all the license notices in the Document, and any Warranty Disclaimers, provided that you also include the original English version of this License and the original versions of those notices and disclaimers. In case of a disagreement between the translation and the original version of this License or a notice or disclaimer, the original version will prevail. If a section in the Document is Entitled "Acknowledgements", "Dedications", or "History", the requirement (section 4) to Preserve its Title (section 1) will typically require changing the actual title.

9. TERMINATION

You may not copy, modify, sublicense, or distribute the Document except as expressly provided for under this License. Any other attempt to copy, modify, sublicense or distribute the Document is void, and will automatically terminate your rights under this License. However, parties who have received copies, or rights, from you under this License will not have their licenses terminated so long as such parties remain in full compliance.

10. FUTURE REVISIONS OF THIS LICENSE

The Free Software Foundation may publish new, revised versions of the GNU Free Documentation License from time to time. Such new versions will be similar in spirit to the present version, but may differ in detail to address new problems or concerns. See http://www.gnu.org/copyleft/. Each version of the License is given a distinguishing version number. If the Document specifies that a particular numbered version of this License "or any later version" applies to it, you have the option of following the terms and conditions either of that specified version or of any later version that has been published (not as a draft) by the Free Software Foundation. If the Document does not specify a version number of this License, you may choose any version ever published (not as a draft) by the Free Software Foundation. ADDENDUM: How to use this License for your documents To use this License in a document you have written, include a copy of the License in the document and put the following copyright and license notices just after the title page: Copyright (c) YEAR YOUR NAME. Permission is granted to copy, distribute and/or modify this document under the terms of the GNU Free Documentation License, Version 1.2 or any later version published by the Free Software Foundation; with no Invariant Sections, no Front-Cover Texts, and no Back-Cover Texts. A copy of the license is included in the section entitled "GNU Free Documentation License". If you have Invariant Sections, Front-Cover Texts and Back-Cover Texts, replace the "with...Texts." line with this: with the Invariant Sections being LIST THEIR TITLES, with the Front-Cover Texts being LIST, and with the Back-Cover Texts being LIST. If you have Invariant Sections without Cover Texts, or some other combination of the three, merge those two alternatives to suit the situation. If your document contains nontrivial examples of program code, we recommend releasing these examples in parallel under your choice of free software license, such as the GNU General Public License, to permit their use in free software.

Printed by Books on Demand GmbH, Norderstedt / Germany